Physics of Strain Hardening of Structural Steels

Yu.F. Ivanov, V.E. Gromov, N.A. Popova,
Yu.A. Shliarova, M.A. Porfiriev, M.D. Starostenkov, A.P. Semin

Copyright © 2023 by the authors

Published by **Materials Research Forum LLC**
Millersville, PA 17551, USA

Published as part of the book series
Materials Research Foundations
Volume 153 (2023)
ISSN 2471-8890 (Print)
ISSN 2471-8904 (Online)

Print ISBN 978-1-64490-276-9
ePDF ISBN 978-1-64490-277-6

Distributed worldwide by

Materials Research Forum LLC
105 Springdale Lane
Millersville, PA 17551
USA
http://www.mrforum.com

Printed in the United States of America
10 9 8 7 6 5 4 3 2 1

Table of Contents

Introduction

The choice of this or that material for a structure is governed by a range of properties, including the relationship between the strength (yield-point and ultimate strength) and plasticity (relative uniform deformation, total elongation per unit length to failure) as well as the fracture toughness and other characteristics.

Metallic, ceramic, polymeric and composite materials make up the majority of structural materials. Metallic materials possess the best relationship between strength and plasticity as compared to that of other structural materials.

These advantages of metallic structural materials have resulted in the fact that the fraction of steel volume among all structural materials exceeds 90%. An improvement in the strength properties of structural materials over recent decades has been caused mainly by the development of alloys having new chemical and phase compositions.

The long-term and stable development of ferrous metallurgy at the worldwide scale depends upon the successful solution of the problem of widening the consumption of metal products. Many international organizations are now actively engaged in projects in this field. The most prospective areas of application of steel products are considered to be machine-building and construction. Special importance is attached to the building industry where new technologies, above all in individual construction, constitute a still larger application of steel structures.

In machine-building the creation of new crystalline materials, coatings and strengthening layers results in the optimization of designs, increases in their reliability, energy- and resource-saving and improvement of the tribological, wear-resistant and strength properties of products. The development of materials possessing the improved physical and mechanical properties has substantial importance in creating new products in space, electro-technical and medical equipment.

Steel still dominates, as before, the balance of consumption of structural materials in most of the developed countries of the world. A decreasing consumption has now already been seen in the form of products of a low added value. The production and consumption of high-quality rolled-steel stock grow moreover at priority rates. The production and consumption of corrosion-resistant steel grow at the same time.

In spite of the progress made in the development of science and technology during the XXth century, mankind continues to live in the era where the major structural material is steel: alloys based upon iron. The 'iron age' of the early history of humanity has already persisted for three thousand years. Although synthetic, polymeric and composite materials have found numerous applications, an alternative to steel for the manufacture of machine parts and structures is nevertheless lacking. This is due to the inimitable physical and mechanical characteristics of steel. The problem of the strength of steel products is closely related to the problems of economical consumption of steels. The materials science of steels therefore continues to develop intensively.

The materials science of steel began to develop intensively in the XIXth century and continues to develop at an ever-increasing rate. By the middle of the XXth century the physical science of steel had begun to develop due to the efforts of the school of academician G.V.Kurdyumov. The application of X-ray structural research methods and the technique of transmission electron microscopy played a major role in many investigations. That is, the application of these methods permitted the detailed structures of steels to be studied and classified. A large contribution to the study of steels was made by the Russian scientists, V.G.Kurdyumov, L.M.Utevskii, V.M.Schastlivtsev, A.M.Glezer, V.I.Izotov, M.A.Shtremel, E.V.Kozlov, M.L.Bernshtein, M.E.Blanter, V.V.Rybin, L.I.Tushinsky, A.A.Bataev and others, by the Ukrainian scientists, Yu.V.Milman, V.N.Gridnev, M.V.Belous, Yu.Ya.Meshkov and V.G.Gavrilyuk, and by scientists of other countries: V.Pitch, E.C.Bain, J.Thomas, A.R.Marder, Ts.Nishiyama, F.B.Pickering, R.L.Fleischer, P.B.Hirsch, G.Kraus and others.

Much success was had in the materials science of steels with regard to establishing the fundamental basis of their strength. Nevertheless a number of important problems in the physics of steel were not developed with due care. In this connection, it is necessary to note the clearly insufficient attention paid to the dislocation structures of steels and their evolution during the course of deformation. This particularly concerned the quantitative parameters of dislocation ensembles. Little attention was paid to fragmentation processes. The internal stress fields were studied mainly by means of X-ray structural analysis, and scant attention was paid to the analysis of morphology and to the classification of bainite and martensite existing at differing temperature intervals.

Comprehensive consideration of the properties characterizing brittle-fracture resistance revealed the possibility of an effective strain hardening of steels of various classes. It is important to understand the prospective fields of application of a technology based upon plastic deformation following heat-treatment and to choose the most appropriate process flowsheet for the deformation treatment of any particular steel. It is necessary to study the dependence of a strengthening effect upon the material's structural state before deformation, and the parameters of the treatment regime, and identify the cause-effect relationships between the phenomena underlying the complexity of property improvement. A knowledge of the deformation mechanisms of a steel structure and its properties under plastic deformation is necessary in order to control the process of strain hardening. This primarily concerns steels having a bainite, martensite and pearlite structure.

The possibilities of increasing the operational and mechanical properties of steel based upon a pearlite structure are already close to exhaustion. Because of this, the attention of researchers in the field of materials science is concentrated on the search for possible routes toward the creation and use of high-strength steels having a bainite structure. The high strength of such structures is due to the fact that the α-phase bainite crystals have a small size and a high dislocation density. The bainite steels are of a new type which simultaneously offers high tensile strength, increased impact strength and good weldability, Because of this, high operational characteristics and a comparatively low primary structural cost are ensured. The steels are called a structural material for the XXIst century. They have already become real competitors to conventional ferrite-pearlite and martensite heat-hardenable steels for two main reasons. Firstly, modern industry needs structures offering high operational characteristics for a relatively low initial cost. With the rapid development of society and its economy, the requirements of steels are constantly growing: the future tall buildings, large engineering structures, large-span bridges, oil pipelines of large dimension exposed to extreme Arctic conditions, large ships, etc. All of these require the use of materials having increased strength, lengthened service life, safety and moreover good weldability as compared to those of conventional steels.

Secondly, the manufacture of bainite steel harmoniously combines the latest breakthroughs and technological achievements in metallurgy and advances in the field of materials science.

The relevance of investigations into the formation of martensite, pearlite and bainite structures in steel, and their evolution under heating and deformation during operation is determined largely by the practical importance of the uses of these steels. The production and application of steels having a different structure should be based upon establishing the physical nature, mechanisms and regularities of their formation and upon the evolution of structural phases, the defect sub-structure and the mechanical behaviour under deformation.

Chapter 1. Evolution of hardened steel structures under uniaxial compression

1.1. Strain hardening curves of hardened steel

It is now clear that a stage-like behavior characterizes the deformation curves [1–14] of both mono- and polycrystals. The behaviour has been studied in most detail for monocrystals of fcc metals and alloys. The characteristics of the various stages of deformation have been distinguished, and studied quite thoroughly. In this section, the results of studying the σ–ε dependence of 38CrNi3MoV (Russian STANDARD GOST 4543-2016 (mass% - C – 0.383; Cr – 1.5; Ni – 3.0; Mo – 0.45; V – 0.18) steel, hardened at differing austenitization temperatures are presented. On the basis of an analysis of the results a stage-like character of the deformation curves is detected. It has been shown, that regardless of the austenitization temperature (α grain size of the initial austenite), the deformation curve of the hardened steel turns out to be a two-stage one.

1.1.1. Strain hardening curves

Deformation of the hardened steel was performed by the uniaxial compression of 4 x 4 x 6mm³ samples at room temperature using an Instron-type machine at a rate of $10^{-2}s^{-1}$, and automatic recording of the load and length-change. The experimental results were subjected to statistical data processing; in this case a value, with a reliable probability of 0.25, did not deviate from <σ> by more than 20MPa, and was taken to be a reliable average value of the stress, <σ>. The Student criterion for the normal distribution law was used when choosing the minimum necessary number of samples required to estimate the average to a given accuracy.

During the compression of samples, especially under large degrees of deformation, the friction forces become substantial at the face surfaces. A graphite lubricant as well as filter-paper spacers impregnated with paraffin were used to decrease these forces. Compression as a deformation technique was convenient to use because, in this case, greater deformations than those possible in tension could be reached. The machine curves of deformation in load (P) versus total elongation (Δl) coordinates were re-calculated and re-plotted as true stress σ versus true deformation ε. The coefficient of strain hardening was also determined as $\theta = \frac{\partial \sigma}{\partial \varepsilon}$.

A characteristic view of the strain hardening curves of 38CrNi3MoV steel is shown in Fig. 1.1. Mathematical processing of the strain hardening curves shows that the σ–ε dependence has a parabolic form and is described by a fourth-power polynomial. Regardless of the austenitization temperature, ranging from 950 to 1200°C, there is thus a clear similarity in the deformation behavior of the hardened steel. In this research the evolution of the steel's structural-phase state was monitored in samples which had been austenitized at 950°C. All further results will therefore concern that material alone.

Materials Research Forum LLC
https://doi.org/10.21741/9781644902776

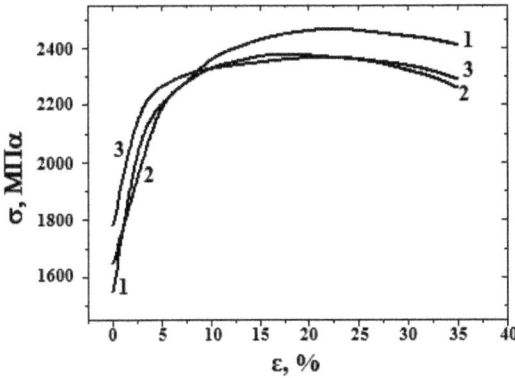

Fig. 1.1. *Strain hardening curves of hardened 38CrNi3MoV steel. Hardening temperature: 1 – 1200 ℃; 2 – 950 ℃; 3 – 1050 ℃.*

1.1.2. Stages of plastic deformation of hardened structural steel

As a rule, the strain hardening of steel is characterized by the strain hardening coefficient $\theta = \frac{\partial \sigma}{\partial \varepsilon}$ determined by differentiation of the σ-ε curve. Upon analyzing the results presented in Fig. 1.2., two stages of strain hardening can be distinguished: a stage with a σ-ε parabolic dependence or a decreasing coefficient of hardening Θ, and a stage with a weakly changing and low value of the coefficient of hardening. If we compare the overall features of the σ-ε and Θ-ε dependences, with those which are observed at these stages in fcc alloys (where the stage-like nature of the flow curves has already been well-studied), then the above-mentioned dependences should be called stages III and IV. In fact, the parabolic σ-ε dependence, the rapid decrease of the hardening coefficient and the band sub-structure is characteristic of stage III. Similar changes in the mechanical characteristics also take place in the steel under study in the present monograph (Fig. 1.2), and the structure of lath martensite is similar to a banded sub-structure in many respects [15]. A constant low hardening and the development of a sub-structure with continuous and discrete misorientations, or a fragmented sub-structure, is characteristic of stage IV. The mechanical characteristics of the steel under study, this time parallel those taking place in fcc-alloys. Fracture of the hardened steel during testing occurred at ε = ~0.27, via brittle cleavage at an angle of ~45° to the deformation axis, to form coarse fragments.

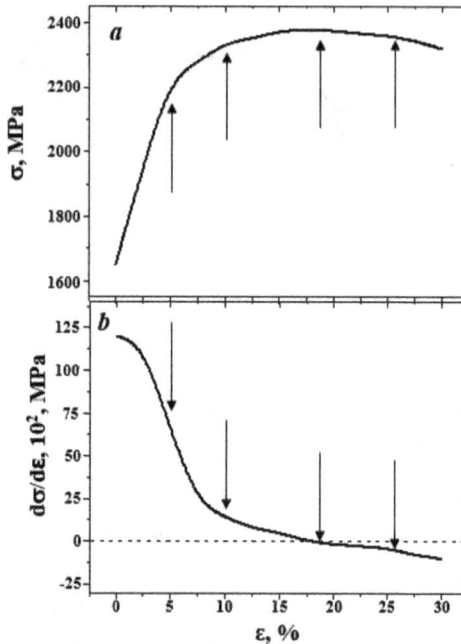

Fig. 1.2. *Strain hardening curve (a) and a strain hardening coefficient as a function of the degree of strain (b) of hardened 38CrNi3MoV steel (T= 950 °C, 1.5h). The arrows indicate the degrees of strain at which electron microscopic studies of the steel's structural-phase state were performed.*

1.2. Evolution of hardened steel structures during straining

Martensite is the basis of a high-strength state in the majority of structural steels. It allows one to exploit various mechanisms of strengthening and softening and create new structural states offering the possibility of obtaining almost any combination of mechanical properties [16–26] during subsequent heat-treatment and/or deformation. There consequently arises the need to analyze comprehensively the phase composition, morphology, and the state of the martensite defect sub-structure which is formed during steel hardening.

In order to foresee the prospective fields of application of technology based upon plastic deformation following hardening, and to choose the most appropriate of strain-treatment flowsheet, the dependence of the hardening effect upon the material's structural state before straining and the parameters of the treatment regime are studied. The cause-effect relationship which determines a combined improvement in properties is also established.

Materials Research Forum LLC
https://doi.org/10.21741/9781644902776

In turn, a knowledge of the regularities of structure formation and steel properties under plastic deformation in the hardened state is necessary in order to control the strain state.

The present chapter analyzes, at the quantitative and qualitative levels, the results obtained when studying the structural-phase state formed as a result of hardening and the evolution of a defect sub-structure and the phase composition of a hardened medium-carbon low-alloy structural steel when subjected to plastic deformation by uniaxial compression.

1.2.1. Structural-phase state of hardened steels before straining

As a rule, a medium-carbon low-alloy structural steel hardened in oil is a multi-phase material and consists of α-phase (a solid solution based upon the bcc lattice of iron), γ-phase (a solid solution based upon the fcc lattice of iron) and carbide particles not dissolved by steel austenitization and/or formed during cooling while hardening (the 'self-tempering' of steel) [16, 19, 21, 26]. A major component of the given class of steel is the α-phase; the volume fraction of retained austenite varies about limits of 10%, and that of carbide-phase particles about a limit of 1%.

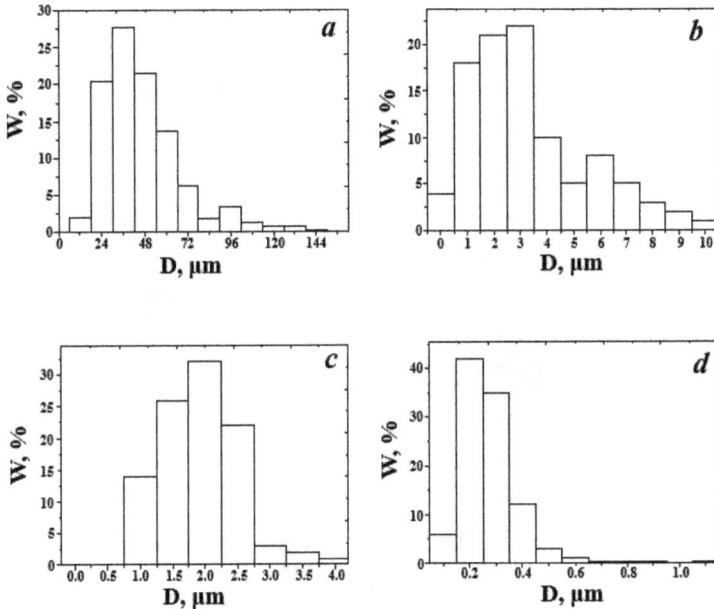

Fig. 1.3. Size distributions of grains (a) packets (b) crystals of lamellar high-temperature martensite (c) and lath martensite (d) formed in 38CrNi3MoV steel subjected to austenitization for 1.5h at 940 °C; b-d are transverse sizes.

Structure of the α-phase

Grain structure of the α-phase. Austenitization of the 38CrNi3MoV steel resulted in the formation of a polycrystalline structure with an average grain size of D = 48.2 ± 23.1μm. The value of the average grain size is often a formal characteristic of a material. The real state of a grain ensemble reflects the size distribution of grains. Fig. 1.3*a* shows the grain size distribution of steel austenitized at 940°C. It follows from an analysis of these results that the histogram is a single-mode one and can be described using logarithms with a normal distribution function. The grain size varies between the limits of 5μm and 153μm. In this case the average size is not the most probable one.

Intragranular structure. Hardening of the 38CrNi3MoV steel leads to the martensite γ ⇒ α transformation. The martensite being formed in a given steel, as shown in [27–30], presents as two morphological components – a lath dislocation martensite (Fig. 1.4*a*) and a lamellar high-temperature (dislocation) martensite (Fig. 1.4*b*). The relationship between the volume fractions of lath and lamellar martensite depends upon the steel austenitization temperature. As the latter temperature is increased, the volume fraction of lamellar high-temperature martensite increases [30–33]. In steel hardened from 940°C the ratio of the volume fractions of packets and crystals of the lamellar high-temperature martensite is 5:1.

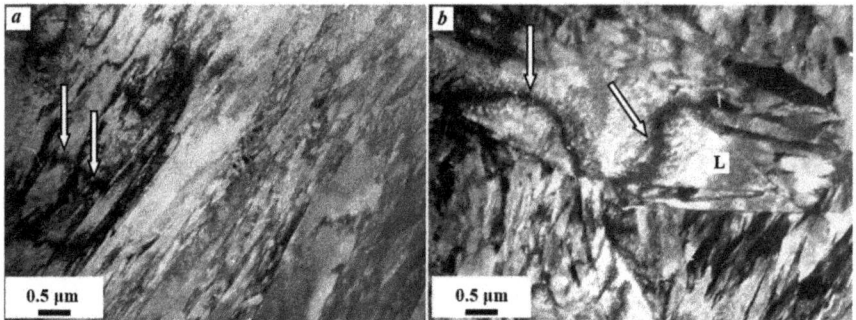

Fig. 1.4. *Structure of hardened steel; a – lath martensite; b – lamellar (L) high-temperature martensite. The arrows indicate bend extinction contours.*

The sizes of the packets and lath martensite crystals increases steadily with increasing grain size of the initial austenite [30–33]. Concurrently with this, the dimensional irregularity of the lath martensite appreciably increases. The transverse sizes of the lath martensite crystals may vary from one tenth to several micrometres. However, the most commonly occurring crystals are those with a transverse size within in the limits of 0.15 and 0.25μm [31–35] regardless of the austenitization temperature. The distribution of the transverse sizes of the martensite crystals obeys a natural logarithmic law [31, 33].

Quantitative analysis of 38CrNi3MoV steel hardened from 940°C showed that the average sizes of the martensite components were: transverse and longitudinal sizes of the packets, 3.8μm and 9μm, respectively, those of the lath martensite crystals, 0.18μm and those of crystals of the lamellar high-temperature martensite, 1.8μm and 12μm, respectively. Analysis of the size distributions of the packets and crystals of lath and lamellar martensite, presented in Fig. 1.3*b-d*, indicates the high degree of dimensional irregularity of the intragranular structure of steel.

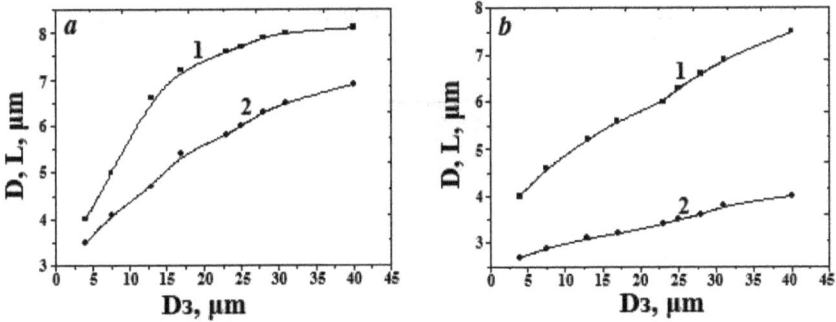

Fig. 1.5. *Average transverse (D) and longitudinal (L) sizes of the packets (a) and plates (b) as a function of the grain size, D₃, for 38CrNi3MoV steel after austenitization at 940 ℃.*

As a rule, the average sizes of the grains, packets and martensite crystals are related to the austenitization temperature of the steel [16, 19, 26, 37]. The scientific reports [30, 31, 33] present results demonstrating the size dependence of the packets and martensite crystals upon the grain size of the initial austenite in which they are located, in the case of 38CrNi3MoV steel. The results of an electron microscopic analysis of the martensite structure of specific grains of steel, hardened from 940°C, using the foil method, are considered below. Fig. 1.5 depicts the size dependence of the packets and the crystals of lamellar high-temperature martensite upon the size of the grains in which they are located. It is clearly seen that, with increasing grain size, the average sizes of both the packets (Fig. 1.5*a*) and the lamellar martensite crystals (Fig. 1.5*b*) increase. Concurrently with this, the dimensional irregularity of the lath martensite and plates increases. The steel austenitization parameters (temperature and time) therefore determine the characteristics of the steel grain structure (average sizes and degree of variation in grain size). The grain structure, in turn, determines the average size and dimensional regularity of the martensite.

Investigation of the structure of specific grains enabled us to define a correlative dependence between the grain size and the grain area occupied by the packet. In determining the grain area (S_3) the latter was approximated by a circumference and the approximation of the packets was done using ellipses. From the results presented in Fig. 1.6 it follows that, with decreasing grain size, the fraction of the area belonging to one packet increases. In the limit of small grains ($D_3 \rightarrow 0$) the real situation is that a grain may be occupied by just one or two packets. Note that the results predicted by us, using this method of approximation, were experimentally verified in [37] by a study of the structure of carbon steel processed by laser. In [37] it was shown that the volume of a small ($\approx 5\mu m$) grain was often occupied by just one packet. Similar results were obtained when studying steel structures processed using a high-current electron beam, i.e. heated and cooled at ultra-high rates [35].

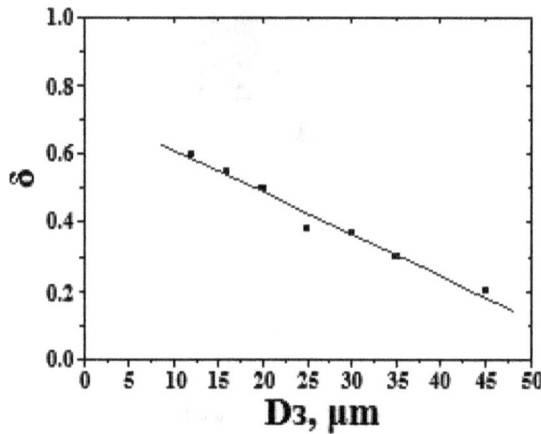

Fig. 1.6. Fraction of the grain area of the initial austenite occupied by the packet ($\delta = S_p/S_3$) as a function of the grain size.

The sub-structure of martensite crystals. Electron microscopic structural examination of crystals of both lath and lamellar martensite detected the presence of a net-like dislocation sub-structure within them. The scalar dislocation density (also known as the *number* dislocation density) reaches a value of $\sim 1.6 \times 10^{11} cm^{-2}$ in the lath martensite crystals and $\sim 1.2 \times 10^{11} cm^{-2}$ in the lamellar martensite crystals [36]. Together with the dislocation sub-structure within the lath martensite crystals, transformation micro-twins in the form of occasional colonies were located along the crystal boundaries (Fig. 1.7) in some cases. Micro-twins were not observed in the lamellar high-temperature martensite.

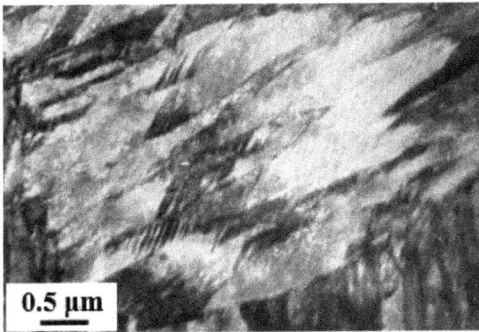

Fig. 1.7. *Transformation twins in the lath martensite crystals of hardened 38CrNi3MoV steel.*

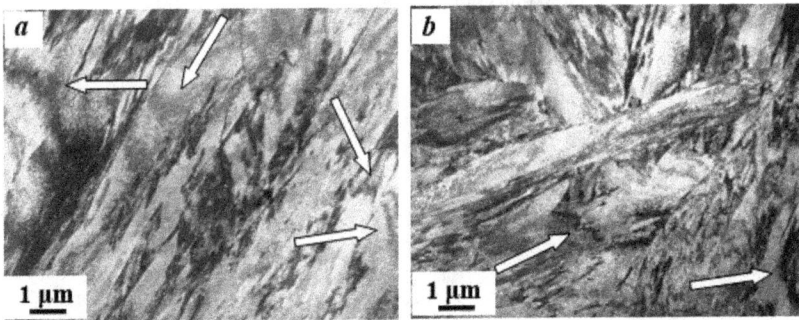

Fig. 1.8. *Bend extinction contours observed in the martensite of hardened 38CrNi3MoV steel.*

Electron microscopic studies of thin foils of the hardened steel resulted in the detection of bend extinction contours in the martensite crystal lattice of the latter (Fig. 1.8). In the lamellar martensite, the bend contours begin and terminate at the boundaries of a plate (Fig. 1.8b); in the lath martensite the contours may intersect a packet, passing from one martensite crystal to another (Fig. 1.8a)

This circumstance allows us to estimate the radial component of the overall misorientation angle β of the martensite crystals in a packet. In scientific reports [34–36] it was shown that, for the martensitic steels, a jump in the extinction contour in transitioning the crystal interface by a distance equal to an intrinsic width corresponded to $\beta \sim 1$ degree. Estimates made using these assumptions showed that β varied within the limits of 1 to 5°. The azimuthal component of the overall misorientation angle α_{az} could

Materials Research Forum LLC
https://doi.org/10.21741/9781644902776

be estimated for a reflection diffusion of the α-phase in an electron-microscopic diffraction pattern, based upon the relationhip, $\alpha_{az} = \frac{\Delta l}{R}$ (where Δl is the width of a tension bar and R is the radius-vector of a reflection) [38]. For a packet, the estimates give values of α_{az} ~1 to 4°. The overall misorientation of the martensite crystals in the packet $\alpha = \sqrt{\alpha_{az}^2 + \beta^2}$ is therefore less than 6.5°. Thus in 38CrNi3MoV steel hardened from 940°C, packets with a low-angle misorientation between the martensite crystals [32, 33] are present together with packets having a high-angle misorientation of the martensite crystals

Retained austenite (the γ-phase)

Quenching of medium-carbon low-alloy steels into water or oil is accompanied by an incomplete γ ⇒ α martensite transformation, i.e. retained austenite [16, 26, 39] is present in the hardened steel structure. In the present steel the retained austenite is seen as thin interlayers (50-70nm) located along crystal boundaries; mainly of the lath martensite. The volume fraction of the retained austenite amounts to 0.05-0.07 of the steel volume (Fig. 1.9).

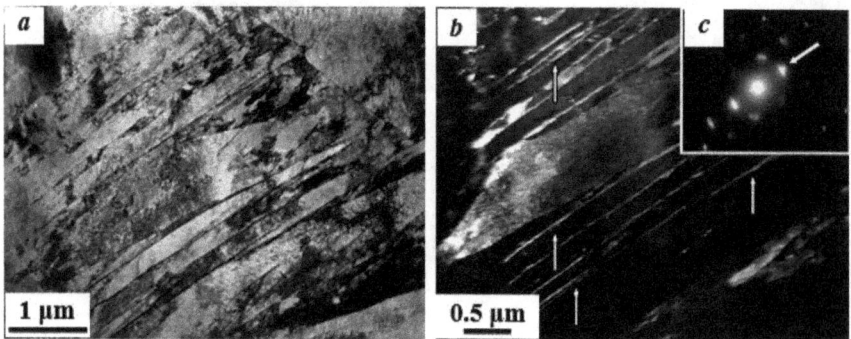

Fig. 1.9. *Interlayers of retained martensite located at the boundaries of the martensite crystals of hardened 38CrNi3MoV steel; a – light field; b – dark field obtained for the [002] γ-Fe reflection; c – electron-microscopic diffraction pattern. The arrows (in b) indicate the retained austenite interlayers; (c) – reflection for dark field production.*

The carbide phase

One of the characteristic features of the martensite hardened low- and medium-carbon low-alloy steels is their instability with regard to decay of the carbon-supersaturated solid solution in the α-iron, with the formation of carbide-phase particles during the cooling process. This phenomenon has been termed a 'self-tempering' of the steel [16, 39]. In the present steel, the carbon-phase particles are found within the bulk and along the boundaries of the martensite crystals [34, 40–42]. In the former case they have needle/plate shapes; in the latter case they take the form of thin interlayers (Fig. 1.10). A micro-diffraction phase analysis has shown that the carbide phase is cementite (Fe_3C). The average sizes of the cementite particles located within the bulk of martensite crystals are: transverse ones, $D \sim 7.6nm$, longitudinal ones, $L \sim 85nm$ with a volume fraction of ~0.3%. The sizes of cementite particles located along crystal boundaries are: transverse ones, $D \sim 50nm$, longitudinal ones, $L = 100$-$500nm$ with a volume fraction of ~0.45%. The cementite particles located along the boundaries of martensite crystals resemble the retained austenite interlayers in shape and position. This fact led to the suggestion that the given morphological variety of self-tempering cementite formed as a result of repeated transformation of retained austenite interlayers during cooling from the martensite initiation temperature to room temperature.

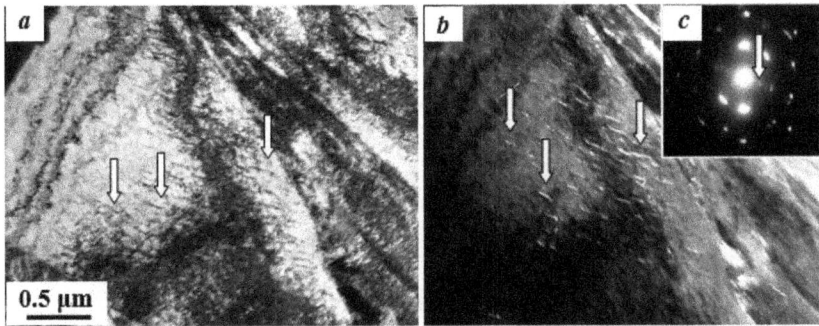

Fig. 1.10. *Cementite particles in the martensite crystals of hardened 38CrNi3MoV steel; a – light field, b – dark field obtained for the [001] Fe_3C reflection; c – electron-microscopic diffraction pattern. The arrows in (a) and (b) indicate the cementite particles; in (c) – the reflection for which a dark field is obtained.*

As a result of hardening a multi-phase morphologically multi-plane structure thus forms in the 38CrNi3MoV steel. By means diffraction electron microscopy of thin foils and X-ray structural analysis, the α-phase and carbide phase (cementite of 'self-tempering') have been detected. It is shown that the α-phase is represented by the lath martensite and the lamellar high-temperature martensite. The structure of individual grains was studied and a relationship was found which connected the size of austenite grains, packets and martensite crystals. By morphological analysis of the cementite formed during martensite cooling it has been shown that the 'self-tempering' of steel is accompanied by two processes. These are decay of the supersaturated solid solution based upon α-iron to form cementite particles of acicular (lamellar) morphology within the bulk of the martensite crystals, and a polymorphic $\gamma \rightarrow \alpha$ transformation of the retained austenite to create cementite particles in the form of interlayers located along the boundaries of the martensite crystals.

1.2.2. Evolution of the defect sub-structure of the martensite of hardened steel during deformation

Deformation of a hardened steel up to fracture of the sample results in no change to the grain structure. Moreover, the steel intragranular structure essentially transforms. To a lesser degree it affects the interfaces of packets and martensite crystals. By means of electron microscopy the boundaries of laths, plates and packets (Fig. 1.11) are clearly detected to be in a fractured state. In some instances however we succeeded in identifying regions of the material where the interfaces of the martensite crystals were not revealed in either light field (Fig. 1.12a) or dark field (Fig. 1.12b) images. This means that the plastic deformation of steel is accompanied by fracture of the interfaces of martensite crystals, packets and plates.

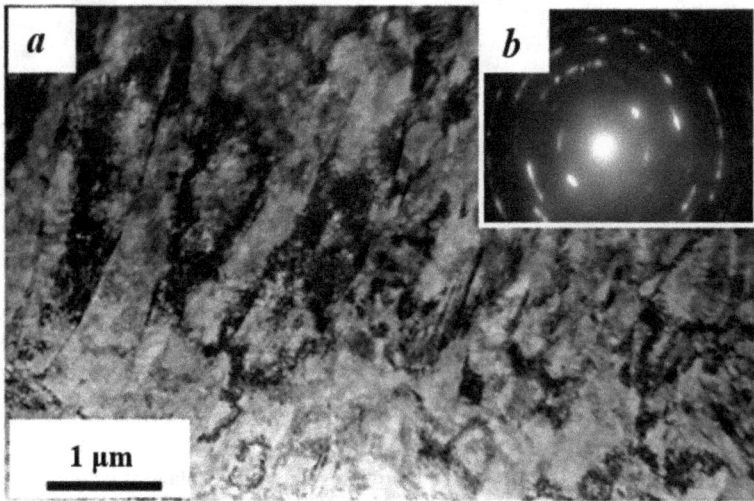

Fig. 1.11. *Martensite structure of hardened 38CrNi3MoV steel, ε = 26% (fractured) a – light field; b – electron-microscopic diffraction pattern.*

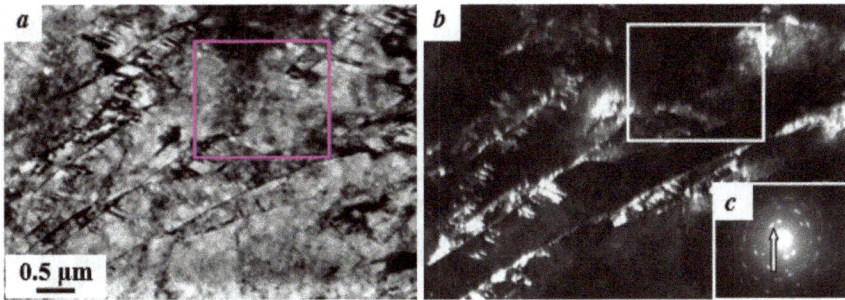

Fig. 1.12. *Martensite structure of hardened 38CrNi3MoV steel, ε = 26% (fractured); a - light field, b - dark field obtained for the [110] α-Fe reflection; c - electron-microscopic diffraction pattern. In (a) and (b) the portion of the structure with fractured boundaries of martensite crystals is singled out; in (c) the arrow indicates the reflection for which a dark field is obtained.*

The sub-structure of the martensite crystals varies substantially with increasing degree of deformation. Firstly, the longitudinal dimensions of the fragments decrease and their degree of misorientation increases. Secondly, the number of micro-twins in the martensite crystals increases (Fig. 1.13). This indicates the occurrence of steel deformation not only via the motion of dislocations but also by twinning. If, in the initial state, micro-twins are observed in separate martensite packets then, in the structure of the fractured steel, the micro-twins are often discovered in several packets which are located side-by-side and in crystals of the lamellar high-temperature martensite.

Thirdly, the morphology of the bend extinction contours changes. That is, in the initial state the contours closed on crystals boundaries, indicating that the sources of stress fields were the intraphase boundaries (Fig. 1.4; Fig. 1.8). Deformation results in the appearance of loop-shaped contours i.e. the contours are located within the packets (Fig. 1.14*a*) or the martensite crystals (Fig. 1.14*b*) and close on themselves. This indicates the operation of stress-field sources within the martensite crystals.

Materials Research Forum LLC

https://doi.org/10.21741/9781644902776

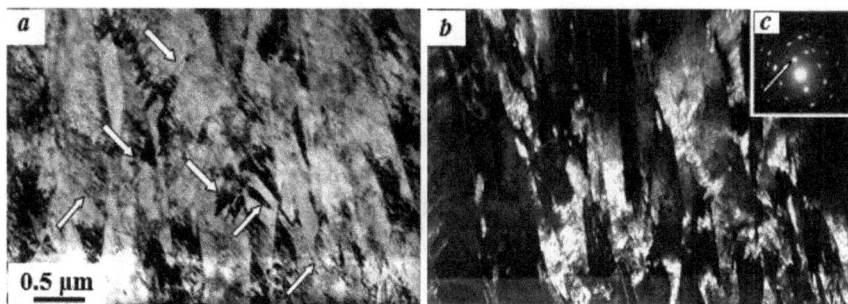

Fig. 1.13. *Deformation micro-twins formed in hardened 38CrNi3MoV steel; ε = 0.26%, a – light field; b – dark field obtained for the [110] reflection of α-Fe; c – electron-microscopic diffraction pattern. In (a) the arrows indicate micro-twins; in (c) the arrow indicates the reflection for which a dark field is obtained.*

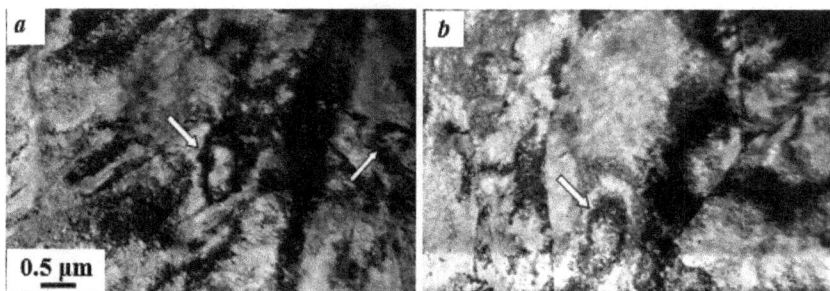

Fig. 1.14. *Structure of hardened 38CrNi3MoV steel formed by plastic deformation (ε = 0.26%); the arrows indicate loop-shaped bend extinction contours.*

Fourthly, deformation is accompanied by the accumulation of continuous and discrete misorientations of sub-structural elements (martensite crystals and crystal fragments). In electron-microscopic diffraction patterns, strands of matrix reflections (Fig. 1.15a,b) are seen which transform into separate point reflections with increasing degree of deformation and form the quasi-circular structure which is characteristic of a non-crystalline material (Fig. 1.15c).

Fifthly, the scalar and excess dislocation densities increase. In this case, the dislocation sub-structure retains and parallels the dense networks. An increase in the degree of deformation up to sample fracture is accompanied by only a decrease in the average dimensions of the dislocation-network elements. The excess dislocation density $\rho_\pm$ is linearly related to the curvature–torsion of the crystal lattice $\chi = b \cdot \rho_\pm$ [12, 13, 43, 44]. The value of χ characterizes the average amplitude of the curvature-torsion of the crystal lattice. Insofar as the overall tensor of the curvature-torsion of the crystal lattice was not measured, a number of tensor components could be deduced from the density of the bend extinction contours. Analysis has shown that the density of bend extinction contours increases with increasing degree of deformation (Fig. 1.16). This indicates the constant growth of a number of tensor components of bending-torsion which differ from zero.

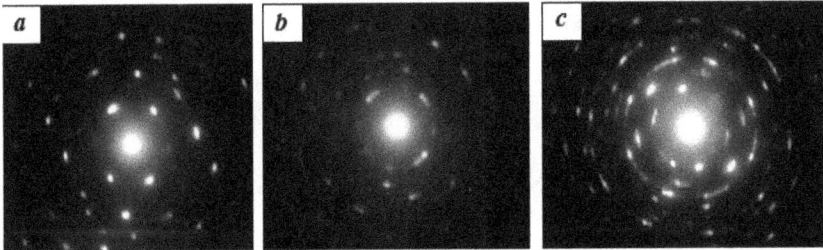

Fig. 1.15. *Characteristic electron-microscopic diffraction patterns obtained from the lath martensite of hardened 38CrNi3MoV steel subjected to plastic deformation with ε = 0.10% (a); 18.6% (b) and 26% (c).*

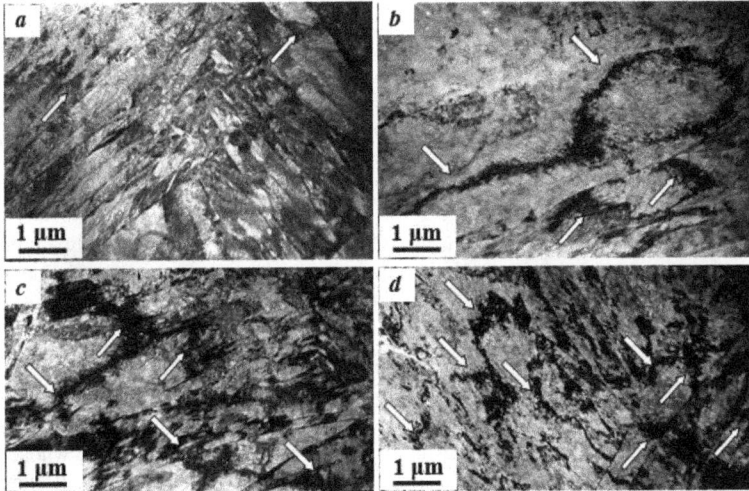

Fig. 1.16. *Structure of hardened 38CrNi3MoV steel formed as a result of plastic deformation to ε = 5% (a); 10% (b); 18.6% (c); and 26% (d); the arrows indicate the bend extinction contours.*

1.2.3. Correlations and regularities of steel structural evolution under deformation

Deformation of the hardened steel is accompanied, as mentioned above, by multiple changes in its structure and phase composition that are reflected by the dependences below.

The deformation is accompanied by a rapid decrease in the longitudinal dimension of the fragments: from 800nm in the initial state to 200nm in the fractured state (Fig. 1.17, curve 1). The transverse size of the fragments is essentially unchanged and is equal to the transvers size of the lath martensite crystals. Note that the fragment sizes (sub-grains) in deformation channels, as shown above, range from 50 to 100nm.

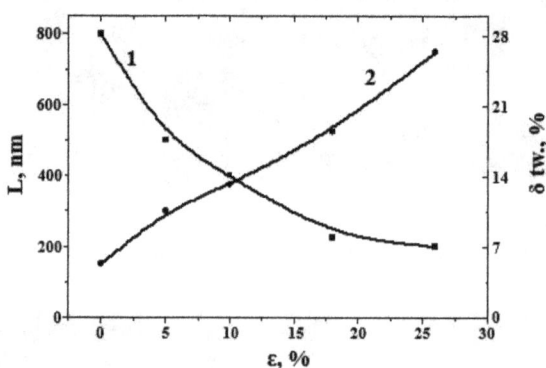

Fig. 1.17. *Longitudinal sizes of fragments of martensite crystals L (curve 1) and the volume fraction δ of twins located within the martensite crystals (curve 2) as a function of the degree of deformation ε of hardened 38CrNi3MoV steel.*

The plastic deformation of the steel is accompanied by deformation micro-twinning. The density of micro-twins increases during deformation but the rate of variation of the characteristic of the steel structure is different at the various stages of hardening (Fig. 1.17, curve 2).

Plastic deformation of the steel is accompanied by an increase in the scalar dislocation density (Fig. 1.18, curve 2). The scalar dislocation density, as mentioned above, cannot completely characterize the material's dislocation sub-system. A further important parameter of a dislocation sub-system is the excess dislocation density $\rho_\pm$, whose value also increases with increasing degree of deformation (Fig. 1.18, curve 1). It should be noted that the range of variation of the scalar and excess dislocation density during deformation of the hardened steel is comparatively small, and this is evidently related to the extremely high values which already exist in the initial state.

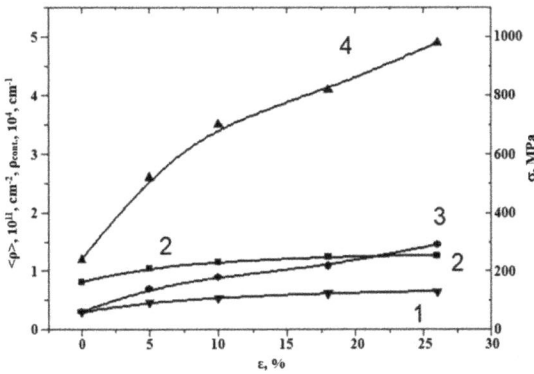

Fig. 1.18. *Excess (curve 1) and scalar (curve 2) dislocation density* $<\rho>$*, linear density of bend extinction contours* $\rho_{cont.}$ *(curve 3) and amplitude of long-range stress* σ *(curve 4) as a function of the degree of deformation* ε *of hardened 38CrNi3MoV steel.*

The presence of the bend extinction contours (Fig. 1.14; Fig. 1.16) in the electron-microscopic images of thin-foil structure indicates a bend-torsion of the crystal lattice, i.e. it reveals the formation of long-range stress fields [43–45] in the steel both upon hardening and during deformation.

Bending of the crystal lattice can be only elastic at first, and is produced by stress fields arising due to incompatibility of the deformation of, for example, the grains of a polycrystal [46, 47] or of plastic material containing non-deformable particles [48]. The sources of the stress fields of elastic origin, that arise mainly under inhomogeneous deformation of the material, are the junctions and grain boundaries of polycrystals [49, 50], dispersed non-deformable particles [48] and, in some cases, cracks [50, 51]. In the second case, bending can be plastic if it is created by dislocation build-up, i.e. an excess dislocation density localized in some volume of the material [12, 13, 44, 46, 47]. In the third case, bending can be elastic-plastic when both sources of stress are present in the material.

The procedure for estimating of the magnitude of the long-range stress fields from the corresponding extinction contours consists of determining the bending-torsion of the crystal lattice [43]. For this purpose, either the velocity of displacement of the extinction contour upon changing the goniometer tilt-angle or the width of the extinction contour is measured. By means of special tests which make simultaneous use of both techniques it has been established that the contour-width in terms of the misorientations in hardened steel is ~1 degree. The curvature-torsion amplitude χ is determined by the value of the continuous misorientation gradient:

$$\chi = \frac{\partial \varphi}{\partial l}, \tag{1.1}$$

where $\partial \varphi$ is the change in orientation of the foil reflecting-plane and ∂l is the displacement of the bend contour.

If dislocations are absent from the portion of the foil being studied, elastic bending-torsion then takes place. The amplitude of the long-torsion stress fields may here be defined by the formula:

$$\sigma_\tau = Gt \frac{\partial \varphi}{\partial l}, \tag{1.2}$$

where G is the shear modulus and t is the foil thickness.

Test-estimates made of hardened steels [15], as well of steels subjected to various degrees and forms of deformation [43, 52], showed that reasonable estimates of the long-range stress fields could be made by using the approximate formula:

$$\sigma_\tau = Gt \frac{\partial \varphi}{\partial l} \approx G \frac{t}{h}, \tag{1.3}$$

where h is the transverse size of the bend extinction contour.

Plastic bending-torsion is provided by a local excess dislocation density: $\rho_\pm = \rho_+ - \rho_-$. In this case, the relationship is [43, 52]:

$$\rho_\pm = \frac{1}{b}\frac{\partial \varphi}{\partial l} = \frac{\chi}{b}. \tag{1.4}$$

In the case of plastic bending-torsion the scalar dislocation density ρ alone should be at least no less than the excess density defined by the formula (1, 4). If the scalar dislocation density, measured locally, is less than the value $\rho_\pm$ ($\rho < \rho_\pm$) then elastic-plastic bending of the crystal lattice occurs. In this last case the value $\rho_\pm$ is a conditional one, because it can never exceed ρ.

The amplitude of the long-range stress fields in the case of plastic bending-torsion can be determined from the formula [43]:

$$\sigma_\tau = Gb\sqrt{\rho_\pm}. \tag{1.5}$$

The morphology of bend extinction contours therefore characterizes the gradient of the bending-torsion of the crystal lattice, the transverse size of contours, the degree of bending-torsion of the crystal lattice and the amplitude of long-range stress fields [43]. The results shown in Fig. 1.18 (curve 4) reveal that the long-range stress fields increase throughout steel deformation.

The excess dislocation density is linearly related to the curvature-torsion of the crystal lattice $\chi = b \cdot \rho_\pm$ and is proportional to the amplitude of the long-range stresses σ [43]. The value of χ characterizes the average amplitude of the curvature-torsion of the crystal lattice. As in the present research, the overall tensor of the curvature-torsion of the crystal lattice is not measured, but a number of tensor components can be deduced from the bend extinction contour density. Analysis showed that the density of bend extinction contours increased with increasing degree of steel deformation; remaining beyond saturation in stage IV of strain hardening (Fig. 1.18, curve 3). The latter demonstrates a constant increase in a number of tensor components of the bending-torsion, other than zero.

Plastic deformation of the hardened steel is accompanied by fragmentation, micro-twinning, twisting and curvature of the lattice of martensite crystals. These processes lead to an increase in the azimuthal and radial components of the overall misorientation of the steel volumes. The first component characterizes both the continuous and discrete misorientation while the second component is related only to discrete ones. A concurrent structural transformation of the martensite packet consists of a decrease in the degree of orientation of the martensite crystals in the structure of a packet. In [26] it has been shown that crystals having six different crystallographic orientations are simultaneously present in the structure of a packet in structural steels. Studies performed by us show that the degree of crystallographic orientation of martensite crystals in a packet decreases to 2 to 3 (Fig. 1.19, curve 2) during deformation. This may indicate the unification of separate martensite crystals, which are close in orientation, by fracture of the boundaries dividing them.

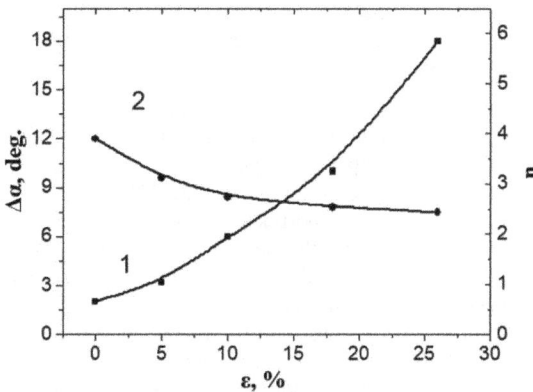

Fig. 1.19. *Azimuthal component of the overall misorientation of the structure by the martensite crystal, Δα (curve 1) and the number of lath orientations in a packet, n, determined by indexing of electron-microscopic diffraction patterns (curve 2) as a function of the degree of deformation, ε, of hardened 38CrNi3MoV steel.*

1.2.4. Deformation channels of hardened structural steel

Detailed analysis of the structure of the deformed state allowed us to reveal the specific material components which were located, as a rule, along intraphase interfaces: that is, the interfaces of neighboring packets or the interfaces of plates and packets – so called channels of localized deformation. In [53, 54] it was shown that under conditions of severe plastic deformation, by drawing at room temperature, of 08Γ2C steel (a ferrite-pearlite structure) a characteristic feature is the formation of extended regions having an ultra-dispersed structure – the channels of deformation. These regions appear as a tabby contrast in dark-field images of matrix reflections. The electron-microscopic diffraction patterns obtained from these regions have, as a rule, a quasi-circular structure. The channels of deformation are the sites of deformation-localization. These regions extend for several tens of micrometres in length and are up to 1μm in cross-section. With increasing degree of deformation, the average size of the deformation channels increases [53, 54]. Within the deformation channels the sub-structure is a fragmentary one, the fragment sizes are however far smaller than in the bulk of the material. The fragments in the deformation channel are moreover isotopic in shape. Judging from the fragment size it can be supposed that a shear which exceeds the average one by several tens of times localizes in the deformation channel. The difference in the forms of the fragments in the matrix (highly anisotropic fragments) and channels (isotropic fragments) is a sure sign of the differing mechanisms involved in their deformation. The isotropy of the fragment shapes in the channel suggests that differing temperature conditions are involved in their formation. If anisotropic fragments are the result of cold deformation, then isotropic fragments are the result of warm deformation.

The next feature of the structure of deformation channels is related to the behaviour of the bend extinction contours within them. Note that the bend extinction contours mark out regions having a similar orientation of specific reflection planes with respect to the incident electron beam [45]. It was established [53, 54] that in both the deformation channel and in regions adjacent to it, portions with a single orientation, or a close one, extended approximately parallel to the length of the channels. By analogy with hydrodynamics, such portions are similar to the lines of flow in a laminar current. Because in drawing operations, a considerable number of portions with a turbulent flow generally arise, such an analogy throws light on the nature of the deformation channels. That is, the conditions of deformation there are such that the work of deformation turns out to be lower than that in neighboring portions. It is assumed that local heating [53, 54] plays the major role here.

One more feature of the deformation channels involves the considerable stress fields which are localized within them and in regions adjacent to them. Two mechanisms of relaxation of the stress fields [53, 54] are observed. One is fragmentation. In this case, chains of fragments of small size and close orientation are formed. These are located along the deformation channel. The other mechanism is the development of micro-cracks.

An example of localized deformation channels being formed during the uniaxial compression of 38CrNi3MoV hardened steel is shown in Fig. 1.20. As mentioned in [53, 54], the deformation channel takes the form of an extended region whose transverse size is ~0.5μm. The deformation channel has a layered structure, resembling that of a martensite packet. The layers consist of crystallites whose sizes range from 50 to 100nm. The circular form of the electron-microscopic diffraction pattern obtained from a region of localization of the deformation channel (Fig. 1.20d) mainly indicates a high-angle misorientation of the crystallites forming it. It is important to note that, in the material regions adjacent to the deformation channel, the structure is similar to that of the initial state according to morphological indicators; i.e. crystals of lath and lamellar martensite are detected. The electron-microscopic diffraction pattern obtained from the foil region adjacent to the channels is the point-like one characteristic of polycrystalline material (Fig. 1.20c). With increasing degree of deformation the material bulk occupied by deformation channels increases; reaching several tens of percent.

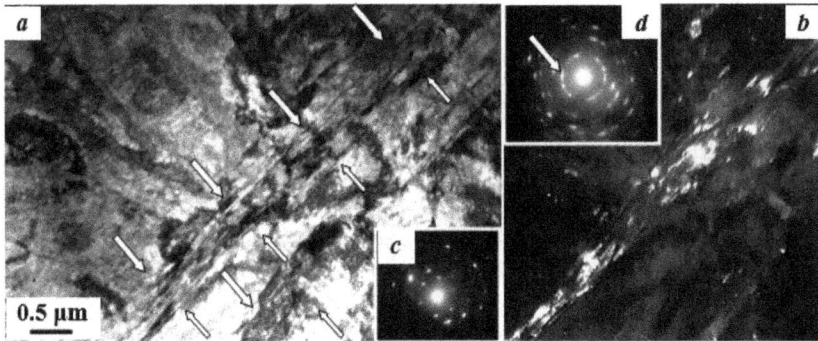

Fig. 1.20. *Deformation channels formed in hardened 38CrNi3MoV steel; ε = 18.6%; a – light field; b – dark field obtained for the [110] α-Fe reflection; c, d – electron-microscopic diffraction patterns. The arrow in (a) indicates the deformation channels; the arrow (d) indicates the reflection for which a dark field is obtained. The electron-microscopic diffraction pattern (c) is obtained from a foil region far from the deformation channel; the electron-microscopic diffraction pattern (d) is obtained from the region of localization of the deformation channel.*

1.3. Evolution of the carbide phase state of hardened steel during deformation; redistribution of the carbon

The results obtained when studying (at the qualitative and quantitative levels) the evolution of a defect sub-structure and the phase composition of a medium-carbon low-alloy hardened structural steel, subjected to plastic deformation, are described here. The average size, density and volume fraction of carbon-phase particles, the volume fraction of retained austenite (γ-phase), the lattice spacing of the α- and γ-phases can be used as parameters which characterize the deformation behaviour and phase composition.

1.3.1. State of the carbide phase 'self-tempering' cementite) before deformation

The carbide phase is the third component of hardened steel (Fig. 1.10). Electron microscopic micro-diffraction analysis showed that the carbides had the composition, Fe_3C; i.e. cementite. The cementite particles were located predominantly within the bulk of the martensite crystals and the lath and lamellar morphology, and took the form of thin plates (needles). The volume fraction of the carbide phase particles was small (0.0075). The reason for their formation was the process of 'self-tempering'; i.e. tempering proceeding within the temperature range between martensite transformation onset, and ambient [16, 39–42].

The steel under study is thus, in the hardened state, a three-phase material; the major phase is martensite; mainly of lath morphology. Electron microscopic micro-diffraction studies enable us to follow the evolution of the γ-phase (retained austenite) and the carbide phase (cementite of 'self-tempering') during deformation.

1.3.2. Evolution of the phase composition of hardened steel during deformation

Deformation of the hardened steel is accompanied by phase transformations. It firstly concerns the retained austenite. With increasing degree of deformation, the volume fraction of the retained austenite decreases noticeably and, at $\varepsilon = 0.2$, retained austenite is no longer detectable using electron-diffraction microscopy of X-ray structural analysis. Cementite nano-dimensional particles (Fig. 1.21) are meanwhile observed along the boundaries of the martensite crystals and packets. A strain transformation of the retained austenite thus proceeds according to:

$$\gamma\text{-Fe} \Rightarrow \alpha\text{-Fe} + Fe_3C$$

Comparing the results obtained by studying the structure and morphology of the localized deformation channels (extended interlayers located along the boundaries of packets and plates) there is an increase in the volume fraction with increasing degree of deformation. Combining this with the fact that there is a decrease in the volume fraction of retained austenite with increasing degree of deformation, it can be surmised that the probable sites of nucleation of deformation channels in hardened steel may be the retained austenite interlayers.

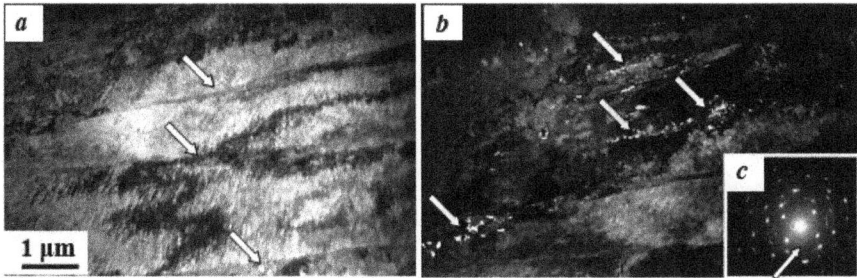

Fig. 1.21. *Structure of hardened 38CrNi3MoV steel formed as a result of plastic deformation (ε = 18.6%); a – light field; b - dark field obtained for the [121] Fe₃C reflection; c – electron-microscopic diffraction pattern. In (a) and (b) the arrow indicates cementite particles; in (c) the arrow indicates the reflection for which a dark field is obtained.*

Evolution of the 'self-tempering' cementite state. Deformation of the hardened steel is also accompanied by the dissolution of cementite particles. This leads to a decrease in the average size, density and volume fraction of the carbide-phase particles. At the same time, the morphology of the particles changes: the initial needle-shaped particles (those located within the bulk of the martensite crystals) and thin interlayers (particles located at the boundaries of martensite crystals and packets) transform into ellipsoidal ones (Fig. 1.22) in the last stage of deformation. These facts indicate a dissolution (fracture) of the 'self-tempering' cementite particles during deformation.

As a rule, two fracture mechanisms of the iron carbide particles in steel under deformation [43, 55, 56] are considered. Firstly, there is the mechanism of cutting of particles by moving dislocations and the passage of carbon atoms into the ferrite matrix. In this case the degree of decay of the cementite particles amounts to some tens of percent of the number of fractured particles. As a rule, cutting of the cementite particles is accompanied by their subsequent distribution within the bulk of the crystal. Secondly, there is the mechanism of dissolution of the particles with the movement of carbon to dislocations (a pulling of carbon atoms from the carbide phase). The latter mechanism is possible due to the appreciable difference in binding energy of carbon atoms to dislocations (~0.6eV) and to iron atoms in the cementite crystal lattice (~0.4eV). Carbon diffusion then proceeds within the stress-field produced by a dislocation-sub-structure around the cementite particles. The degree of cementite decay is determined by the value of the dislocation density and the type of dislocation sub-structure. For example, cold-drawing of Ст70 steel to ε = 16% results in a concentration of up to 0.13% carbon at dislocations. It is evident that both fracture processes of cementite particles take place in the material under study. A sharp decrease in the coefficient of anisotropy of the particles in the initial stage of deformation indicates, by all appearances, a predominant role being played by the fragmentation of particles by cutting at this stage.

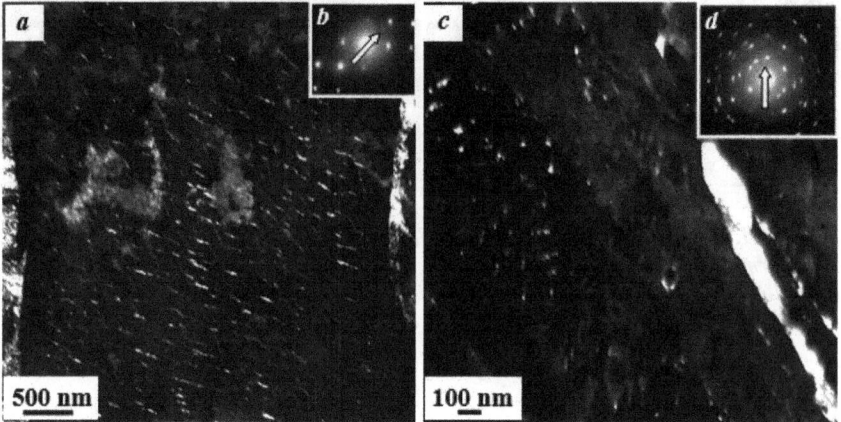

Fig. 1.22. *Structure of hardened 38CrNi3MoV steel formed by plastic deformation; a, b –*
ε = 10%; c, d – ε = 26%; a, c – dark fields obtained for reflections [111] Fe₃C (a) and
[102] Fe₃C (c); b, d – electron-microscopic diffraction pattern. In (b) and (d) the arrows
indicate the reflection for which a dark field is obtained.

As discussed above, steel deformation results in the fracture of cementite particles. The latter is characterized by a decrease in the average size, linear density (Fig. 1.23) and volume fraction (Fig. 1.24) of carbide-phase particles located within the bulk and at the boundaries of martensite crystals. The carbon atoms are released from the crystal lattice of the steel crystal structure (interphase and intraphase boundaries, dislocations). This should lead to an increase in the steel's lattice parameter with increasing degree of deformation. As studies involving X-ray structural analysis have shown, the steel's crystal lattice parameters indeed increase abruptly in the initial stage ($\varepsilon = 5\%$) of deformation and later reach saturation (Fig. 1.25).

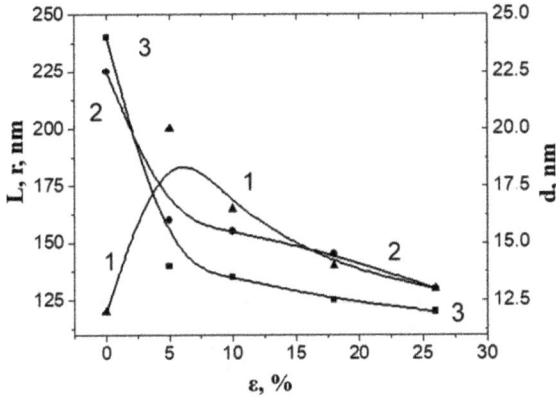

Fig. 1.23. *Transverse d (curve 1) and longitudinal L (curve 2) sizes of the cementite particles located in the bulk of cementite crystals and the distance between the cementite particles r (curve 3) as a function of the degree of deformation ε of hardened 38CrNi3MoV steel.*

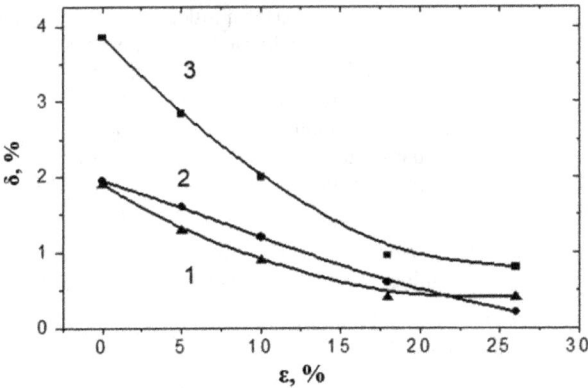

Fig. 1.24. *Volume fraction of cementite particles δ located within martensite crystals (curve 1), at the boundaries of martensite crystals (curve 2) and their sum (curve 3) as a function of the degree of deformation ε of hardened 38CrNi3MoV steel.*

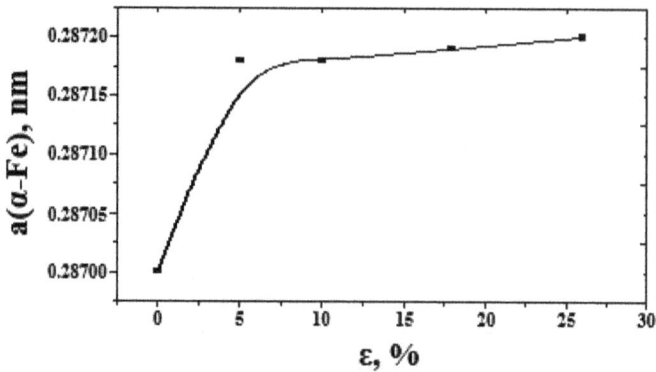

Fig. 1.25. *The α-Fe lattice parameter as a function of the degree of deformation ε of hardened 38CrNi3MoV steel.*

1.3.3. Redistribution of carbon during steel deformation

The observed quantitative regularities of variations in the parameters of steel structures during plastic deformation allowed us to perform studies which revealed the possible sites of carbon atoms in the structure of deformed steel. For this purpose the carbon distribution was determined as a function of the degree of deformation. In [56–65] it was shown that the carbon in the steel structure could be found in the following places: in the carbide-phase particles, in a solid solution based upon α- or γ-iron and in structural defects (dislocations, sub-boundaries, interphase and intraphase boundaries). Estimates of the amounts of carbon concentrated in the locations listed above were made by using the expressions given in Table 1.1.

Note that the estimate of the amount of carbon concentrated in the cementite crystal lattice was performed by assuming a stoichiometric composition for the phase.

The results of the estimations of the amounts concentrated at various sites in the steel structure are tabulated in Table 1.2.

Table 1.1. Analysis technique of carbon distribution in steel

Site of carbon location	Expressions for estimation	References
Solid solution based upon α-iron	$\Delta C_\alpha = \Delta V_\alpha \dfrac{a_\alpha - a_\alpha^0}{39 \pm 4} \cdot 10^3$, ΔV_α – volume fraction of α-phase; a_α – current lattice parameter, $a_\alpha^0 = 0.28668\ nm$	[61–63]
Solid solution based upon γ-iron	$\Delta C_\gamma = \Delta V_\gamma \dfrac{a_\gamma - a_\gamma^0}{44} \cdot 10^3$, ΔV_γ – volume fraction of γ-phase; a_γ – current lattice parameter, $a_\gamma^0 = 0.3555\ nm$	[64, 65]
Cementite particles	$\Delta C_k = 0.07 \cdot \Delta V_k$, ΔV_k – volume fraction of carbide phase	[57]
Structural defects	$\Delta C_{def.} = C_0 - (\Delta C_\alpha + \Delta C_\gamma + \Delta C_k)$, $C_0 = 0.38wt\%$ – carbon content in steel according to nomenclature	[66]

Table 1.2. Distribution of carbon in the structure of hardened steel subjected to plastic deformation

ε, %	ΔC in crystal lattice, wt.%					ΔC, wt%, total	ΔC, wt%, on defects
	α-Fe	γ-Fe	Fe₃C (at boundaries)	Fe₃C, in matrix	Fe₃C, total		
0	0.1125	0.048	0.07	0.1243	0.1943	0.3548	0.0252
5	0.1350	0.016	0.07	0.1143	0.1843	0.3353	0.0447
10	0.1375	0.016	0.0612	0.105	0.1662	0.3197	0.0603
18.6	0.1425	0.008	0.0613	0.091	0.1523	0.3028	0.0772
26	0.1425	0.0	0.063	0.049	0.112	0.2545	0.1255

These estimates show that, as the degree of deformation increases, the total amount of carbon located in solid solutions based upon α- and γ-iron decreases, and the amount of carbon located at structural defects increases. The process of carbon migration to defects is especially intensive at $\varepsilon \geq 0.2$.

The results obtained when studying deformed samples which were subjected to subsequent tempering at 50 to 70°C became convincing proof of the carbon distribution illustrated above. In samples which fractured under deformation (ε = 0.26), carbide-phase particles located along dislocation lines (Fig. 1.26) were detected following tempering. Estimates made from electron microscopic images of steel structure have shown that the sizes of the particles range from 2.0 to 2.5nm. Electron-microscopic diffraction patterns obtained from these precipitates reveal diffusion strands to be located at the sites of carbide-phase reflections (Fig. 1.26). This may reflect, firstly, 'floating' parameters of the crystal lattice of the precipitated particles, indicating a variable composition of the particles, secondly, a defective structure for the particles and, thirdly, their small size.

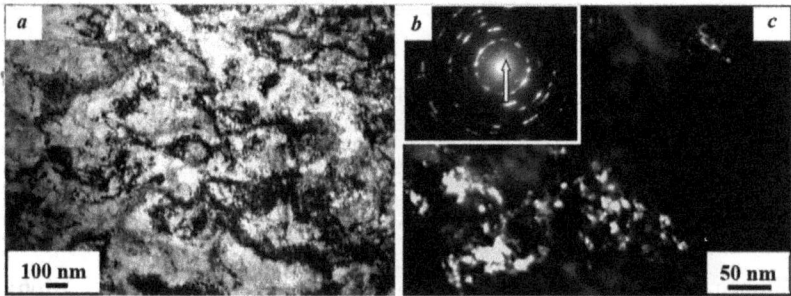

Fig. 1.26. *Structure formed in pre-hardened 38CrNi3MoV steel subjected to plastic deformation (ε = 26%) and subsequent tempering at 60 ℃; a – light field; b – electron-microscopic diffraction pattern; c – dark field obtained for the [110] Fe₃C reflection. In (b) the arrow indicates the reflection for which a dark field is obtained.*

Chapter 1 Conclusions

Analysis of the results presented in this chapter allows us to draw the following conclusions:

1. Austenitization at 950°C (for 1.5 hours) and subsequent oil-hardening of the 38CrNi3MoV steel results in the formation of a multi-phase material whose major phase is martensite; mainly of lath morphology.

2. Studies of the hardened steel, subjected to plastic deformation by uniaxial compression, using electron diffraction microscopy and X-ray structural analysis, revealed a complex interrelated evolution of the phase composition and defect sub-structure of the material. This manifests itself at the macro (sample as a whole, structure of the grain ensemble), meso (packets, martensite crystals, retained austenite), micro (defect sub-structure of martensite crystals, carbon-phase particles) and nano (redistribution of carbon atoms upon fracture of carbide-phase particles) structural levels.

3. It is shown that the deformation of the hardened steel is accompanied by dislocation glide and micro-twinning;

4. It is established that, as the degree of deformation increases, a decrease in the longitudinal size of martensite crystal fragments is observed.

5. The formation of channels of localized deformation: particular material states located along the interfaces of neighboring packets and interfaces of plates and packets is detected.

6. A variation in the phase composition of the steel during deformation, caused by repeated transformation of the retained austenite, is revealed.

7. It is established that deformation of the hardened steel is accompanied by the fracture of cementite particles; the freed carbon atoms entering a solid solution based upon α-iron and also defects of the crystal lattice.

Chapter 1 References

[1] Sachs, G., Weerts, J., Die Verfestigungskurven. Kupfer, Silber, Gold, Z.Physik, 1930, 62, 473-481. https://doi.org/10.1007/BF01339674

[2] Stepanov, A.V., Die plastischen Eigenschaften der Silberchlorid- und Natriumchlorid-Einkristalle, Phys. Z. Sowjetunion, 1935, 8[1] 25-40.

[3] Seeger, A., Mechanism of gliding and hardening in cubic face-centred and hexagonal close-packed metals, Dislocation and mechanical properties of crystals, IIL, 1969, 179-289.

[4] Jaoul, B., Gonzalez, D., Deformation plastique de monocristaux de fer, J.Mech. Phys. Sol., 1961, 9, 16-38. https://doi.org/10.1016/0022-5096(61)90036-9

[5] McLean, D., Mechanical Properties of Metals, Metallurgiya, 1965, 431pp.

[6] Ivanova, V.S., Ermishkin, V.A., Strength and Plasticity of Refractory Metals, Monocrystals, Metallurgiya, 1975, 80pp.

[7] Pavlov, V.A., Physical Fundamentals of Cold Deformation of BCC Metals, Nauka, 1978, 208pp.

[8] Vasillieva, A.G., Strain Hardening of Quenched Structural Steels, Mashinostoenie, 1981, 321pp.

[9] Bell, J.F., Experimental Basis of Mechanics of Deformed Solids - Part II, Nauka, 1984, 431pp.

[10] Koneva, N.A., Lychagin, D.V., Zhukovsky, S.P., Kozlov, E.V., Evolution of dislocation structure and stages of plastic flow of polycrystalline iron-nickel alloy, Physics of Metals and Metal Science, 1985, 60[1] 171-179.

[11] Trefilov, V.I., Moiseev, V.F., Pechkovski, E.P. et al, Strain Hardening and Fracture of Polycrystalline Metals (Ed. V.I.Trefilov), Naukova Dumka, Kiev, 1981, 248pp.

[12] Koneva, N.A., Kozlov, E.V., Physical nature of phase character of plastic deformation, Structural levels of plastic deformation and fracture (Ed. V.E.Panin),

Materials Research Forum LLC
https://doi.org/10.21741/9781644902776

Nauka, Siberian branch, Novosibirsk, 1990, 123-186.

[13] Koneva, N.A., Kozlov, E.V., Physics of substructural hardening, Bulletin TSUAB, 1999, 1, 21-35.

[14] Berner, R., Kronmüller, G., Plastic Deformation of Monocrystals, Mir, 1969, 272pp.

[15] Kozlov, E.V., Popova, N.A., Ivanov, Yu.F., Teplyakova, L.A., Band substructure and lath martensite structure. Comparison of evolution ways, Proceedings of Higher Schools - Physics, 1992, 10, 13-19. https://doi.org/10.1007/BF00559881

[16] Kurdyumov, V.G., Utevskii, L.M., Entin, R.I., Transformations in Iron and Steel, Nauka, 1977, 236pp.

[17] Gudremon, E., Special Steels, V. I, II. (Translated from German), Metallurgiya, 1966, 1274pp.

[18] Meskin, V.S., Fundamentals of Steel Alloying, Metallurgiya, 1964, 684pp.

[19] Lysak, L.I., Nikolin, B.I., Physical Fundamentals of Thermal Treatment of Steel, Tekhnika, Kiev, 1975, 304pp.

[20] Petrov, Yu. N., Defects and Diffusionless Transformation in Steel, Naukova Dumka, Kiev, 1978, 267pp.

[21] Blanter, M.E., Phase Transformations during Heat Treatment of Steel, Metallurgiya, 1962, 268pp.

[22] Novikov, I.I., Theory of Metal Thermal Treatment, Metallurgiya, 1978, 392pp.

[23] Lakhtin, V.M., Physical Metallurgy and Heat Treatment of Metals, Metallurgiya, 1977, 407pp.

[24] Gulyaev, A.P., Metal Science, Metallurgiya, 1978, 647pp.

[25] Pickering, F.B., Physical Metallurgy and Treatment of Steel, Metallurgiya, 1982, 184pp.

[26] Schaslivtsev, V.M., Mirzaev, D.A., Yakovleva, I.L., Structure of Heat Treated Steel, Metallurgiya, 1994, 288pp.

[27] Ivanov, Yu.F., Kozlov, E.V., Electron-microscopic analysis of the 38CrNi3MoV steel martensite phase, Proceedings of High Schools - Ferrous Metallurgy, 1991, 8, 38-41.

[28] Ivanov, Yu.F., Kozlov, E.V., Morphology of martensite phase in low-and medium-carbon low-alloy steels, Thermal Treatment and Physics of Metals, 1990, 15, 27-34.

[29] Ivanov, Yu.F., Kozlov, E.V., Investigation into the effect of cooling rate on structural parameters of the 38CrNi3MoV steel, Proceedings of High Schools - Ferrous Metallurgy, 1991, 6, 50-51.

[30] Ivanov, Yu.F., Kozlov, E.V., Examination of effect of austenitization parameters on

Materials Research Forum LLC
https://doi.org/10.21741/9781644902776

the morphology of the martensite phase of 38CrNi3MoV steel, Physical Metallurgy and Metal Science, 1991, 11, 202-205.

[31] Ivanov, Yu.F., Effect of technological parameters on the dimensional homogeneity of lath martensite, Physical Metallurgy and Metal Science, 1992, 9, 57-63.

[32] Ivanov, Yu.F., Effect of alloying degree of material on the lath martensite structure of iron alloys and steels, Proceedings of Higher Schools - Ferrous Metallurgy, 1995, 10, 52-54.

[33] Ivanov, Yu.F., Influence of grain size of initial austenite on lath martensite structure of iron alloys and steels, Proceedings of Higher Schools - Ferrous Metallurgy, 1995, 12, 33-38.

[34] Ivanov, Yu.F., Kozlov, E.V., Multi-stage diagram of martensite transformation of low- and medium-carbon low-alloy steels, Materials Science, 2000, 11, 33-37.

[35] Ivanov, Yu.F., Kozlov, E.V., Bulk and surface hardening of structural steel - morphological structural analysis, Proceedings of Higher Schools - Physics, 2002, 45[3] 5-23. https://doi.org/10.1023/A:1020384414722

[36] Ivanov, Yu.F., Investigation into dislocation substructure of crystals of martensite phase of structural steels subjected to different regimes of thermal treatment, Transaction "Evolution of Dislocation Substructure, Hardening and Fracture of Alloys", TGU, Tomsk, 1992, 52-59.

[37] Bernshtein, M.L., Kaputkina, L.M., Prokoshkin, S.D., Tempering of Steel, MISIS, 1997, 336pp.

[38] Utevskii, L.M., Diffraction Electron Microscopy in Physical Metallurgy, Metallurgiya, 1973, 584pp.

[39] Thomas, J., Phase transformations and microstructure of alloys with high hardness and fracture toughness. Possibilities and limitations of their use in development of alloys, Problems of Development of Structural Alloys, Metallurgiya, 1980, 176-203.

[40] Ivanov, Yu.F., Kozlov, E.V., Self-tempering of steel - analysis of kinetics of carbide-formation processes, Proceedings of Higher Schools - Ferrous Metallurgy, 1990, 12, 38-40.

[41] Ivanov, Yu.F., Kozlov, E.V., Morphology of cementite in the martensite phase of 38CrNi3MoV steel, Physical Metallurgy and Metal Science, 1991, 10, 203-204.

[42] Ivanov, Yu.F., Kozlov, E.V., Analysis of carbide-formation kinetics during self-tempering and low-temperature tempering of structural steel, Transaction "Defects of Crystal Lattices and Properties of Metals and Alloys", TulPI, Tula, 1992, 90-94.

[43] Gromov, V.E., Kozlov, E.V., Bazaikin, V.I., Tsellermayer, V.Ya., Ivanov, Yu.F. et al. Physics and Mechanics of Drawing and Die Stamping, Nedra, 1997, 293pp.

[44] Koneva, N.A., Kozlov, E.V, Nature of substructural hardening, Proceedings of Higher Schools - Physics, 1982, 8, 3-14. https://doi.org/10.1007/BF00895238

[45] Hirsch, P.B., Howie, A., Nicholson, R.B. et al. Electron Microscopy of Thin Crystals, Mir, 1968, 577pp.

[46] Panin, V.E., Likhachev, V.A., Grinyaev, Yu.V., Structural Levels of Deformation of Solids, Nauka, Novosibirsk, 1985, 229pp.

[47] Pybin, V.V., Large Plastic Deformations and Metal Fracture, Metallurgiya, 1986, 224pp.

[48] Ashelby, J., Continuum Theory of Dislocations, IIL, 1963, 247pp.

[49] Vladimirov, V.I., Physical Theory of Hardness and Plasticity. Point Defects, Strengthening and Recovery, LPI, 1975, 120pp.

[50] Shtremel, M.A., Strength of Alloys - Part I. Defects of Lattice, MISIS, 1999, 384pp.

[51] Finkel, V.M., Physical Principles of Fracture Deceleration, Metallurgiya, 1977, 359pp.

[52] Koneva, N.A., Kozlov, E.V., Trishkina, L.I., Lychagin, D.V., Long-range stress fields, curvature-torsion of crystal lattice and stages of plastic deformation. Methods of measurement and results, New Methods in Physics and Mechanics of Deformed Solid. Transactions of International Conference, TGU, Tomsk, 1990, 83-93.

[53] Gromov, V.E., Kozlov, E.V., Panin, V.E., Ivanov, Yu.F. et al. Channels of deformation under conditions of electrostatic stimulation, Metallophysics, 1991, 13[11] 9-13.

[54] Ivanov, Yu.F., Gromov, V.E., Kozlov, T.V., Sosnin, O.V., Evolution of localized deformation channels in a process of electrostimulated drawing of low-carbon steel, Proceedings of Higher Schools - Ferrous Metallurgy, 1997, 6, 42-45.

[55] Tushinsky, L.I., Bataev, A.A., Tikhomirova, L.B., Structure of Pearlite and Design Strength of Steel, Nauka, Novosibirsk, 1993, 280pp.

[56] Babich, V.K., Gul, Yu.P., Dolzhenkov, I.E., Strain Ageing of Steel, Metallurgiya, 1972, 320pp.

[57] Ivanov, Yu.F., Gladyshev, S.A., Popova, N.A., Kozlov, E.V., Interaction of carbon with defects and processes of carbide formation in structural steels, Transaction "Interaction of Defects of Crystal Lattice and Properties", TulPI, Tula, 1986, 100-105.

[58] Speich, G., Swann, P.R., Yield strength and transformation substructure of quenched iron-nickel alloys, J.Iron and Steel Inst., 1965, 203[4] 480-485.

[59] Raj, B.V.N., Thomas, G., Transmission electron microscopy characterization of dislocated lath martensite, Proc. Int. Conf. Martensite Transformation ICOMAT-

1979, Cambridge, 1979, 1, 12-21.

[60] Speich, G., Tempering of low-carbon martensite, Trans. Met. Soc. AIME., 1969, 245[10] 2553-2564.

[61] Kalich, D., Roberts, E.M., On the distribution of carbon in martensite, Met. Trans., 1971, 2[10] 2783-2790. https://doi.org/10.1007/BF02813252

[62] Fasiska, E.J., Wagenblat, H., Dilatation of alpha-iron by carbon, Trans. Met. Soc. AIME., 1967, 239[11] 1818-1820.

[63] Barnard, S.J., Smith, G.D.W., Sarikaya, M., Thomas, G., Carbon atom distribution in a dual phase steel: an atom probe study, Scripta Met., 1981, 15[4] 387-392. https://doi.org/10.1016/0036-9748(81)90216-7

[64] Ridley, N., Stuart, H., Zwell, L., Lattice parameters of Fe-C austenite at room temperature, Trans. Met. Soc. AIME., 1969, 246[8] 1834-1836.

[65] Veselov, S.I., Spektor, E.Z., Dependence of austenite lattice parameter on carbon content at high temperatures, Physical Metallurgy and Metal Science, 1972, 34[5] 895-896.

[66] Pridantsev, M.V., Davydova, L.N., Tamarina, I.A., Structural Steels (reference book), Metallurgiya, 1980, 388pp.

Physics of Strain Hardening of Structural Steels   Materials Research Forum LLC
Materials Research Foundations **153** (2023)   https://doi.org/10.21741/9781644902776

Chapter 2. Strengthening mechanisms of quenched steel

In increasing the strength characteristics of metallic materials, three fundamental types of strengthening are generally distinguished: 1) solid-solution strengthening (substitutional and interstitial atoms, structural vacancies, short-range and long-range order, antiphase domains, etc.), 2) sub-structural strengthening caused by linear and planar defects and 3) multiphase strengthening (carbides and inclusions of retained martensite in steels, decay of eutectics, composites, etc.). Strengthening by radiation and quenching defects (vacancies, heat and radiation, inherent interstitial atoms, etc.) is subsumed by the previous three types [1–14].

During the past decades, great emphasis has been placed on the quantitative estimation of a steel's physical properties and considerable progress has been made in understanding their mechanical characteristics on the basis of microstructural analysis [1, 7, 9, 12, 14]. Particular attention was focused on strength; a feature which nowadays can often be predicted with sufficient reliability on the basis of the alloy composition and microstructure [1, 14]. The strength is often analyzed using physical models, sometimes based upon empirical and semi-empirical premises; especially when it is necessary to describe properties on the basis of analyses of the complex microstructures formed in steel, such as martensite or bainite.

In order to exploit the strength typical of steels to the full and, at the same time, obtain an optimum combination of the properties needed for their successful application, an understanding of their strengthening mechanisms becomes important. It is moreover necessary to know the factors which control a mechanism and their effect upon many other properties; especially viscosity and plasticity. During the past decades, a number of reviews of the mechanisms of strengthening [1-14] have been published. In this research therefore only the most modern ideas, published in recent years, and having a direct relationship to explaining the properties of steels will be considered.

Following the results shown in [1-14], we shall assume that the principal factors determining the properties of a quenched carbon steel are the following: the availability of a solid solution, the boundaries (sizes) of grains, packets, martensite crystals, second-phase particles (cementite of 'self-tempering'), inclusions of retained austenite and dislocations and internal stress-fields caused by structural elements. Presented below will be numerical estimations of the strengthening mechanisms, based upon the results of quantitative analysis of the parameters of a steel structure in the initial (quenched) state, and as a function of the degree of its deformation. That is, the evolution of the strengthening mechanisms of a quenched structural steel when in the quenched state and at various stages of its deformation.

Based upon the results of examination of the structure of the quenched steel, estimates were made of the contributions of the following mechanisms of retardation of the moving dislocations: retardation due to forest dislocations, cementite particles, intraphase boundaries and to interaction with internal stress-fields. Estimates of the total strength of the steel were made by using additive and square-law (from full-strength obstacles) summations of the contributions [1].

Estimates of the contributions made by various strengthening mechanisms, and of the total strength of the steel, were calculated for different stages of the strain-strengthening of steel (see Chapter 1). This enabled us to perform an evolution analysis of the steel-strengthening mechanisms, and steel strength as a whole, as a function of the degree of deformation.

2.1. Strengthening of quenched structural steel by intraphase boundaries

Up to the present time, two types of boundaries are distinguished: non-cuttable (high-angle boundaries of grains, packets and martensite crystals) and cuttable (low-angle interfaces of sub-grains and martensite crystals) by glide dislocations.

Non-cuttable boundaries. As found experimentally the strength, not only the yield point [15, 16] but also the flow stress in a region of homogeneous deformation [17, 18], is related to the grain size by the Hall-Petch relationship:

$$\sigma_y = \sigma_0 + kd^{-1/2}, \tag{2.1}$$

where σ_0 is the friction stress of a material lattice (stress required for dislocation motion, such as the Peierls stress for pure metals), d is the average grain size and k is a proportionality coefficient. In the literature, three models which explain the dependence of the flow stress upon the grain size [6] are considered in detail: 1) a model which relates the grain size to the concentration of stresses in individual slip-bands (Fig. 2.1), 2) a model of strain-strengthening according to which, as the grain size decreases, the dislocation density increases sharply in separate grains thus contributing to an increase in the flow stress, 3) a model based upon the idea of a governing role being played by surface grain-boundary sources of dislocations during glide transfer from grain to grain.

The coefficient k in equation (2.1) is a characteristic of the boundary state of the given material and depends upon the material's purity with regard to impurities, the degree of material deformation, boundary structure and boundary strengthening by second-phase particles. It is established that metal refining to remove impurities, as well as an increase in the degree of deformation, results in a decrease of the value of k [1, 6, 7, 14].

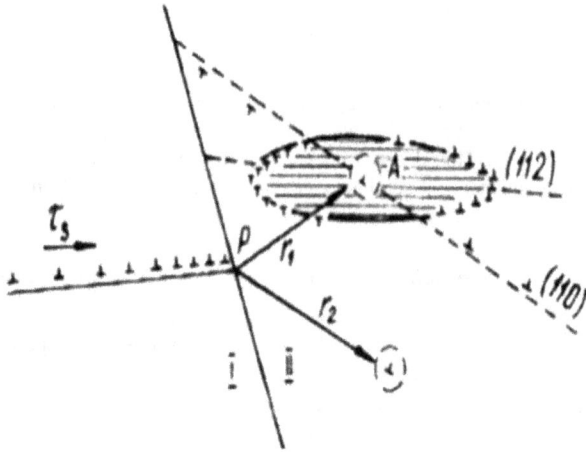

Fig. 2.1. *Schematic diagram of the transfer of plastic deformation across a boundary between grains I and II by gliding and twinning [6].*

As the result of a martensite transformation, the density of intraphase boundaries in steel increases sharply at the expense of the appearance of high-angle boundaries of packets and plates and of the low-angle boundaries of lath martensite crystals. Material strengthening by the high-angle boundaries is estimated using equation (2.1), where d in the quenched steel is the average size of packets or plates and the proportionality coefficient k for boundaries of that type varies from 0 to 5kgf/mm$^{3/2}$ [19, 20].

Cuttable boundaries. Material strengthening by low-angle boundaries (structural strengthening, strengthening by lath boundaries in a packet) is estimated using the expression:

$$\sigma_c = \sigma_0 + k \cdot l^{-m}, \tag{2.2}$$

where $m = 1$ or $\frac{1}{2}$, and l is the effective size of the martensite crystals [21]; being determined by the effective length of a slip plane in the martensite. It is found that, with $m = 1$, the value of k varies from 0.015 to 0.01kgf/ mm$^{3/2}$ while, with $m =1/2$, the value of k varies from 0.2 to 0.98kgf/mm$^{3/2}$ [20, 21].

The first term in the Hall-Petch equation (2.1) σ_0 is, as mentioned above, the friction stress of a material lattice, i.e. the stress required for dislocation motion in pure monocrystals (for instance, the Peierls stress for pure metals) (Fig. 2.2). The stress σ_0 therefore depends greatly upon the degree of material purity and the amount of cold-working. For a theoretically pure material, σ_0 = 17MPa. Experimentally determined values of σ_0 vary from 27 to 60MPa [6, 22]. For steels, values of σ ranging from 30 to 40MPa are normally used [19].

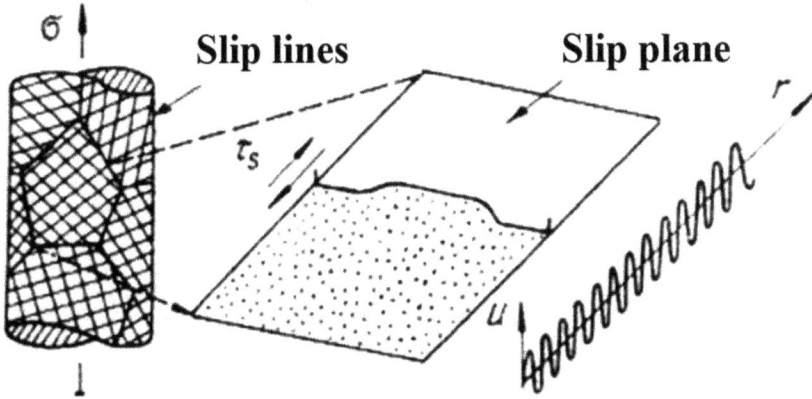

Fig. 2.2. *Deformation determined by dislocation slip and controlled by lattice resistance (the behavior of the force field in a dislocation slip plane is shown on the right) [23].*

Fig. 2.3. shows the total contribution of intraphase boundaries (boundaries of grains, packets, martensite crystals and fragments) to the strain-strengthening of 38CrNi3MoV quenched steel as a function of the degree of deformation. It is clearly seen that, with increasing degree of deformation, the amount of strengthening by intraphase boundaries increases slightly (from 440 to 480MPa), caused by a decrease in the average sizes of the martensite crystals and fragments (see Chapter 1).

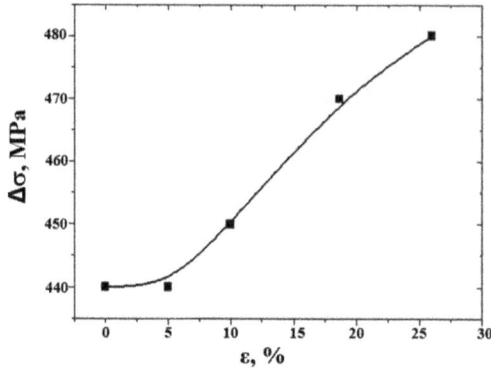

Fig. 2.3. *Contribution of martensite crystal boundaries to the flow stress as a function of the degree of deformation.*

2.2. Dislocation-strengthening of quenched structural steel

The stress required to maintain plastic deformation, i.e. the flow stress σ, is related to the dislocation density by [1, 6–8, 24–28]:

$$\sigma = \sigma_0 + k\sqrt{\rho}, \tag{2.3}$$

where σ_0 is the flow stress of non-dislocation origin (i.e. caused by other strengthening mechanisms), ρ is the average (scalar) dislocation density, $k = m\alpha Gb$, m is the Schmid orientation factor, α is a parameter characterizing the value of interlocation interactions (equal to between 0.1 and 0.51) [22, 29], G is the shear modulus and b is the Burgers vector of the dislocation. For steels, the orientation factor m is usually taken to be such that $m\alpha \approx 0.5$. The relationship (2.3) is actually based upon the principle of similitude of a dislocation structure [26]. In the simplest case (monocrystals of pure well-annealed metals) deformation will be governed by dislocation glide and controlled by a lattice resistance (Fig. 2.2).

From [1, 6] it follows that different sub-structures are characterized by different values of the k parameter. Chaotically distributed dislocations, flat clusters and homogeneous net-like dislocation structures offer the lowest resistance to moving dislocations. The resistance is much higher in ball, cellular, net-like and band sub-structures. A sub-structure with continuous and discrete misorientations occupies an intermediate position with respect to this parameter. Analysis shows that every class of sub-structure determines its own level of glide-dislocation retardation. The basis for it is, in every case, a set of mechanisms; not infrequently with contributions independent even of the scalar dislocation density. A diagram of strain-strengthening of polycrystalline metals with the bcc-lattice, illustrating interaction of the k parameter with the type of dislocation structure in material at various stages of deformation, is shown in Fig. 2.4.

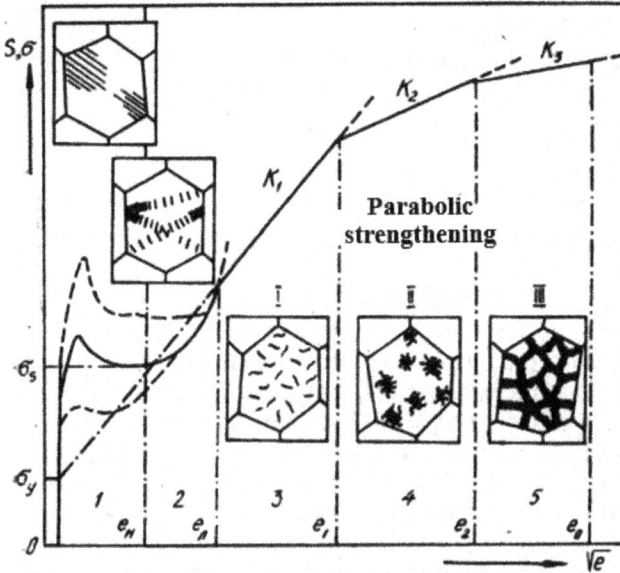

Fig. 2.4. *General diagram of the strain-strengthening of polycrystalline metals with a bcc-lattice: 1 – initial stage; 2 – linear stage; 3-5 – three stages of parabolic strengthening [6].*

It is shown in [8, 30, 31] that the values of the α parameter in different classes of sub-structure increase in the order: chaotic distribution of dislocations $\Rightarrow$ clusters $\Rightarrow$ homogeneous net-like dislocation sub-structure $\Rightarrow$ sub-structure with continuous and discrete misorientations $\Rightarrow$ balls $\Rightarrow$ heterogeneous net-like $\Rightarrow$ cells $\Rightarrow$ band misoriented dislocation sub-structure.

As a rule, glide across sites of average dislocation density does not occur; deformation is realized at 'weak' sites, i.e. parts of the distribution of free dislocation segments which corresponds to their greatest lengths participate in deformation. The picture observed satisfies Kocks criterion [32]: more than 1/3 of dislocation segments should be mobile, for plastic deformation. The formula (2.3) should thus be made more precise by introducing the statistical coefficient χ:

$$\sigma = \sigma_0 + \chi \cdot m \cdot \alpha \cdot G \cdot b \cdot \sqrt{\rho}. \tag{2.4}$$

Kocks [32] predicted a value of χ = 0.84 to 0.87 for similar structures. The value χ detected in practice [8, 30, 31], based upon experimental data even if it depends upon the degree of deformation is on average equal to 0.85.

Fig. 2.5 presents the dependence of the value of the contribution- due to the scalar dislocation density, to the strain-strengthening of 38CrNi3MoV quenched steel upon the degree of deformation. It is clearly seen that, with increasing degree of deformation, the value of the contribution increases insignificantly, ranging from 290 to 360MPa, due to the increase in the scalar density of dislocations located within the bulk of martensite crystals and fragments (see Chapter 1).

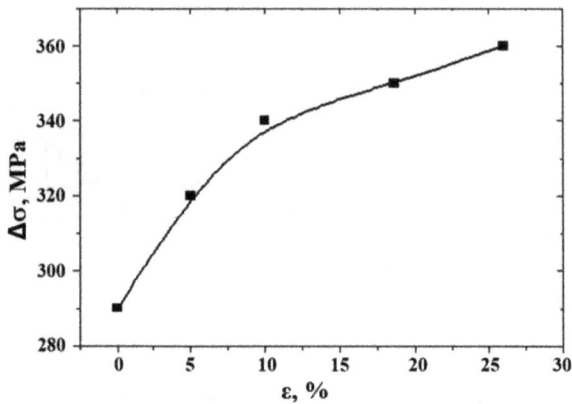

Fig. 2.5. *Contribution to the flow stress arising from 'forest' dislocations as a function of the degree of deformation.*

2.3. Strengthening of quenched structural steel at the expense of internal stresses

An important characteristic with regard to the motion and multiplication of dislocations, and consequently to the analysis of a material's strain-strengthening mechanisms is the field of internal stresses produced by the various material defects [7, 8, 30, 31, 33]. Short-wave (hundreds of angstrom), medium-wave (up to 1μm) and long-wave (1 to 100μm) components of the internal stress fields can be distinguished. The medium- and short-wave components of the internal stress fields play a decisive role in dislocation sub-structures without misorientation (dislocation chaos, dislocation sub-structure). In such sub-structures, as a rule, long-wave components of these fields, caused by polarization of the dislocation sub-structure and the initiation of dislocation-disclination sub-structures [8] are absent.

From the point of view of electron microscopy, the medium- and short-wave components of the internal stress fields are usually deduced from 1) the radius of curvature of free dislocations or dislocation segments (static fields) [34], 2) the distance between dislocations gliding in the same or different planes (elastic fields) [35] and 3) the distance between active glide planes (dynamic fields) [35].

Studies [8, 30, 31] have shown that the medium- and short-wave components of internal stress fields are substantial only at the beginning of plastic deformation. Their magnitude decreases as deformation increases. It follows moreover from analysis of the presented [8, 30, 31] results that the internal stress fields create no strain-strengthening phenomenon in the net-like dislocation sub-structure which is characteristic of the dislocation sub-structure of martensite crystals in quenched steel [36]. The only source of strengthening, whose role increases during deformation, is the contact dislocation retardation by individual barriers (thresholds, reactions, etc.) along dislocation lines; accounted for by relationship 2.4. The retardation is treated as a dislocation 'friction'. The increase in barrier-density along dislocation lines can explain strain-strengthening of the alloy by the formation of a net-like sub-structure within it.

The long-wave (1 to 100μm) components of an internal stress field (hereafter referred to as long-range stress fields) play the most important role in the processes of deformation and strengthening of the quenched steel. Their appearance is related to a polarization of the dislocation structure and to the formation of dislocation-disclination sub-structures [37], to incompatibility of neighboring grains [9, 38], to crystallites of various phases [39] and to presence of micro-cracks [11, 12, 40]. In a foil cut from deformed material, it is possible to measure the retained long-range stress fields caused by operation of the above processes during deformation. The presence of the long-range stress fields results in the bending of a thin foil, or a corresponding curvature of the crystal lattice if a foil retains a plate shape. The possibility of employing electron microscopy for the study of long-range stress fields was recognized by Hirsch co-workers [41]. The method of practically measuring the long-range component of static fields was worked out by E.V.Kozlov and N.A.Koneva [8, 30, 31, 42-44]. The method is original in many respects and can be used in research and industry (the technique for measuring long-range fields was presented in detail in chapter 1).

Fig. 2.6 shows the value of contribution to the flow stress, arising from long-range stress fields, as a function of the degree of deformation of quenched 38CrNi3MoV steel. It is clearly seen that, with increasing degree of deformation, the value of the given contribution increases considerably, rising from 27 to 91kg/mm^2. This is caused by the curvature-torsion increment of the steel crystal lattice due to incompatibility of the deformations of martensite crystals, packets, grains and carbide-phase particles (see Chapter 1).

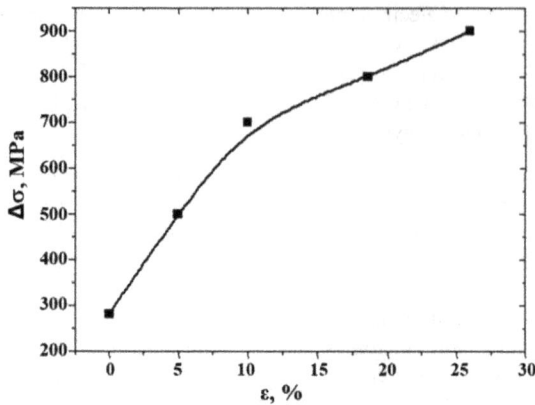

Fig. 2.6. *Contribution to the flow stress arising from internal stress fields as a function of the degree of deformation.*

2.4. Strengthening of quenched structural steel by second-phase particles

Multi-component alloys into which interstitial elements are introduced already have a sufficiently complex structure. The improvement in the strength characteristics of these systems, especially the yield point, as compared with homogeneous materials is caused by the presence of structural irregularities which create an additional resistance to dislocation motion. According to [19, 45] these irregularities can be classified as: 1) local changes caused by fluctuations in composition which lead to the formation of metastable groups/clusters that can exist for a long time at low temperatures due to retarded diffusion processes, 2) metastable volumes such as Guinier-Preston zones (pre-precipitation), 3) second-phase precipitates having a coherent or incoherent interface with the matrix, as well as second-phase inclusions, 4) mixtures of two phases constituting a polycrystal whose composition in separate zones may differ.

In connection with the problems of research the consideration of cases of strengthening by dispersed precipitates and second-phase inclusions is of the greatest interest which, in practice, can be formulated as strengthening by coherent and incoherent particles.

The start of plastic deformation of a crystal is determined by the stress necessary to initiate dislocation motion. For metals its value is very low. The theoretical upper bound on the stress necessary to start the deformation can be determined from the condition that there are no dislocations capable of gliding in the crystal. These conditions are realized only in filamentary crystals (whiskers). For pure iron at 20°C the lower bound is σ = 10MPa while the upper bound is σ_{th} = 10GPa (i.e. 10^3 times higher) [45, 46]. In order to estimate precipitation strengthening we proceed from the fact that dislocations always exist, or may be easily formed, at the beginning of deformation. In metallic materials they may appear most frequently near to grain boundaries. The main reason for precipitation strengthening is the creation of obstacles to dislocation motion in the bulk lattice. Thanks to this the yield point of the material σ_y increases by $\Delta\sigma_H$.

$$\sigma_y = \sigma_0 + \Delta\sigma_H. \qquad (2.5)$$

Where σ_0 is the friction stress in an undistorted lattice while particles that are not cut by mobile dislocations (incoherent particles) act as obstacles to the process of deformation by causing a curvature of the dislocation lines which depends upon the distance between particles and upon the stress (Fig. 2.7).

Fig. 2.7. *Diagram of the by-passing of non-cut particles by dislocations via the Orowan mechanism [75].*

The contribution to material strengthening due to the presence of incoherent particles is estimated to be [46]:

$$\sigma_{op} = M \frac{mG_mb}{2\pi(|\lambda-D|)} \Phi \cdot ln\left(\left|\frac{\lambda-D}{4b}\right|\right), \qquad (2.6)$$

Where D is the average size of the particles, m is an orientation multiplier equal to 2.75 [47] for bcc materials, $\Phi = 1$ for screw and $\Phi = (1-V)^{-1}$ for edge dislocations and M is a parameter which takes account of the non-uniformity of the particle distribution in the matrix and ranges from 0.81 to 85 [48]. There exist several more ways to account for material strengthening by second-phase particles and improve upon the Orowan model. The Fisher, Hart and Pry models [47] posit the formation of dislocation rings around particles which then create additional stresses. Eshelby [49] supplemented the model by proposing that, at large deformations, a relaxation of the stresses arising from dislocation rings takes place upon their departure to secondary glide systems. The Ansell and Janell mechanism [50] considers the fracture of particles under deformation. The Hirsch and Humphreys model [51] accounts for material strengthening at the expense of transverse gliding in interaction with particles.

Alloys which are strengthened by coherent particles behave very differently. The coherent precipitates arise in the early decay stages of supersaturated solid solutions.

Physics of Strain Hardening of Structural Steels
Materials Research Foundations **153** (2023)

Materials Research Forum LLC
https://doi.org/10.21741/9781644902776

Cutting of the particles is observed during the deformation of such materials. When analyzing the dependence of the critical shear stress in a slip plane upon the number of the passing dislocations it might be expected that the coefficient of strengthening would be equal to zero, or negative. The strengthening of alloys with small coherent particles is in fact very modest. After little, or even negative, strengthening under low degrees of deformation, more severe deformation leads to strengthening which occurs as the result of interaction of dislocation clusters from different slip planes. At high degrees of deformation it might be expected that the coherent particles would be refined because of their frequent cutting on different glide systems and would finally dissolve. This is observed in reality. The alloy then exhibits properties which approximate those of a severely deformed homogeneous solid solution.

There are many reasons why the cutting of particles impedes dislocation motion, but all of them result in the creation of a force which counteracts dislocation motion and results in their bending [52]. These are:

1) a stress field around particles because of a difference in the lattice parameters,

2) an ordered arrangement of atoms in coherent particles,

3) a difference in the values of stacking defects; fault energy, elastic constants, Peierls stress,

4) an increase in the surface area of the particle,

5) a transformation occurring within a metastable particle upon the passage of dislocations,

6) transformation of a coherent interface into an incoherent one,

7) defects concentrated on the interface in the case of incomplete coherence between the particle and matrix lattices.

The largest contribution to the resistance to dislocation motion is mainly due to the long-range elastic stress fields near to coherent precipitates which arise as a result of the difference in atomic volumes of the precipitating phase and the solid solution [53-55].

In the presence of zones of pre-precipitates and/or coherent particles in a material, the strengthening is estimated using the mechanism suggested by Mott and Nabarro [56], from which it follows that:

$$\sigma_b = 2G_m \cdot \varepsilon \cdot f, \tag{2.7}$$

where G_m is the matrix shear modulus and f is the volume-fraction of particles,

$$\varepsilon = \frac{3K\delta}{3K+2E(1+v)}, \quad \delta = 2\frac{a_b+a_m}{a_m}.$$

Here δ is the misfit parameter, a_m and a_b are the lattice parameters of the matrix and precipitate, K is the bulk compression modulus of the precipitates and v is the matrix

Poisson ratio. In a given case the strengthening is governed by the volume-fraction of particles and their degree of misfit with the matrix. In other research [56] the additional particle/matrix interfaces and misfit dislocations which appear during the cutting of coherent particles by dislocations were taken into account, resulting in the expression:

$$\sigma_b = 2G_m \cdot \varepsilon^{\frac{3}{2}} \cdot (r_b \cdot f_b b_m^{-1})^2, \tag{2.8}$$

where r_b is the particle radius and b_m is the Burgers vector of the dislocation gliding through the matrix.

In order to estimate the strengthening of a precipitation-hardened alloy it is important to know whether the dislocations by-pass or cut them. For bending around particles, the alloy strengthening is much higher than for cutting (Fig. 2.8).

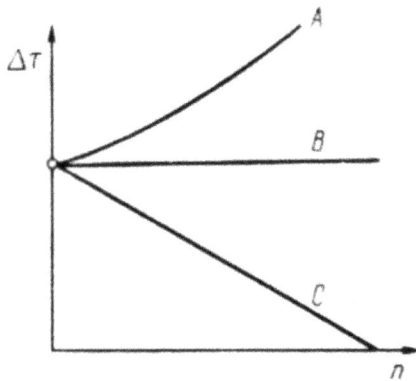

Fig. 2.8. *Variation of the critical shear stress $\Delta\tau$ in a glide plane following the passage of n dislocations; A – by-passing mechanism; B – homogeneous solid solution; C – shearing mechanism [1].*

The reason for this is the fact that, in bending around a dislocation, rings always form and strengthen the glide plane; new dislocations do not arise during cutting. In this case the cross-section of particles decreases and, due to this, their relative effectiveness as obstacles may even decrease (Fig. 2.7). Taking the critical size for the cutting of particles D_{crit} to be [56]:

$$D_{crit} = \frac{4G_m b_m^2}{0.33\pi \cdot b_v G_v}, \tag{2.9}$$

(G_v being the shear modulus of the particle and b_v the Burgers vector of the dislocation moving through the particle) it is found that, for most carbonitride particles precipitated in steel, the critical size is $D_{crit.} \leq 5nm$. Larger particles are not cut by moving dislocations.

Analysis performed during the present research shows that particles of self-tempering cementite are present in quenched 38CrNi3MoV steel. The sizes of the cementite particles before deformation exceed D_{crit}. Estimates of steel strengthening during deformation, taking account of the presence of cementite particles, should therefore be made by using the relationships obtained for incoherent precipitates (Fig. 2.9). Fig. 2.9 shows the value of the contribution made by cementite self-tempering particles as a function of the degree of deformation of quenched 38CrNi3MoV steel. It is clearly seen that, with increasing degree of deformation up to $\varepsilon = 5\%$ the value of the contribution increases and varies from 200 to 300MPa. Under further deformation the contribution of cementite particles to the material strengthening decreases, due to their dissolution (see Chapter 1).

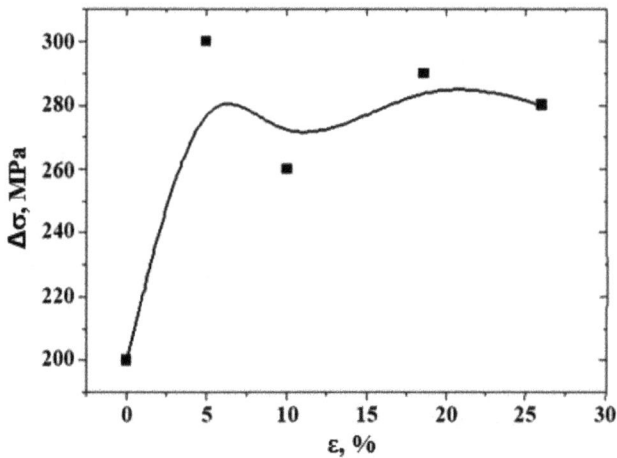

Fig. 2.9. *Contribution of 'self-tempering' cementite particles to the flow stress as a function of the degree of deformation.*

2.5. Strengthening of quenched structural steel by solid-solutions

There are substitutional and interstitial solid solutions. It is known that the introduction of substitutional atoms into iron results in an increase in strength [1, 7, 19, 57, 58]. The degree of strengthening of such a solid solution depends upon a number of factors:

1. The difference in the atomic size of the solvent atoms and of the atoms dissolved in it. The degree of variation in the strength as a function of the concentration of the alloying components is usually determined by the relationship:

$$\frac{d\tau}{dc} \approx \left(\frac{1}{a} \cdot \frac{dc}{da}\right)^n, \tag{2.10}$$

where τ is the shear stress, a is the lattice parameter of the solid solution and c is the concentration of the solute. Data which confirm the influence of the difference in atomic sizes have been obtained for copper alloys [59], ferromagnetic steels [60] and austenitic stainless steels [61].

2. There can be a perturbation of the electron structure, usually expressed in terms of the shear moduli of the solvent atoms and dissolved atoms. In substitutional solutions the dissolved atoms introduce mainly symmetrical lattice distortions of the solvent, and lead to relatively modest strengthening effects. As shown below, the asymmetrical distortions usually caused by interstitial atoms in α-iron result in a far greater strengthening of solid solutions. The intensity of strengthening in substitutional solutions may nevertheless increase substantially if atoms of the solution are connected as dipoles; resulting in an apparent asymmetry of the distortions.

In general the effect of atomic size on the strength of substitutional solutions can be expressed in terms of concentration [1, 7, 19, 57, 58].

$$\sigma \approx C^{1/2} \tag{2.11}$$

where σ is the yield point.

In dilute solid solutions, often the case for a wide range of steels, the relationship above can be simplified by introducing a linear dependence of the degree of strengthening of a solid solution upon the concentration, expressed in atomic percent when the solvent and dissolved atoms do not differ greatly in atomic mass, or as a linear dependence upon concentration in mass percent [1, 7, 19]. By way of illustration data for ferrite are shown in Fig. 2.10 [1, 60].

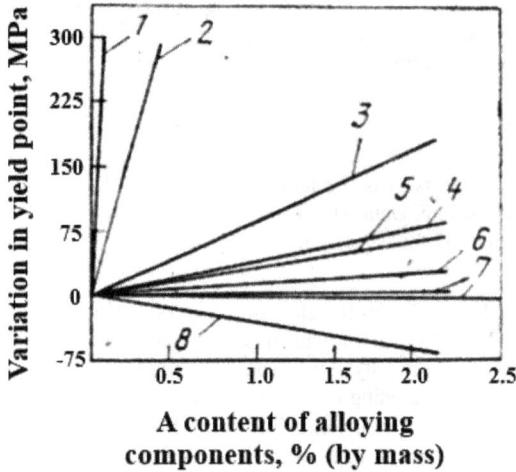

**A content of alloying
components, % (by mass)**

*Fig. 2.10. Effect of alloying on the strengthening of a solid solution in low-carbon ferritic
steels: 1 C+N; 2 P; 3 Si; 4 Cu; 5 Mn; 6 Mo; 7 Ni, Al; 8 Cr content.*

Interstitial solutions. In interstitial solutions such as carbon and nitrogen in α-iron, an asymmetrical lattice distortion occurs, resulting in a far greater strengthening effect as compared to the strengthening of substitutional solutions. A very strong interaction between interstitial atoms in a solid solution and dislocations, which is related to a tendency of dissolved atoms to deposit on dislocations, contributes to this. In this case it can be shown that the strengthening is again proportional to the square root of the solute concentration. Fig. 2.10 presents the dependences of the strength characteristics upon the concentration of alloying elements, from which it is seen that the effect of strengthening in interstitial solid solutions is 10 to 100 times greater than the strengthening effect in substitutional solid solutions [1, 60].

An especially effective example of strengthening in interstitial solid solutions is martensite strengthening. In this case the Fleischer model [62] is acceptable for analyzing the results.

Combined effect of strengthening by the formation of substitutional and interstitial solid solutions. The overall strengthening effect of the formation of substitutional and interstitial solid solutions is usually governed by several factors.

1. The formation of compounds having a limited solubility of interstitial atoms, such as TiC, TiN, etc. In spite of the fact that this leads to a certain decrease in the overall level of strengthening due to solid-solution formation, it can be compensated to some extent by

a strengthening due to the precipitation of second-phase particles. Second-phase particles, being precipitated at grain boundaries, often contribute moreover to their refinement or, in some cases, have an appreciable impact upon the processes of recovery and recrystallization [19].

2. The formation of 'complexes' or bound substitutional and interstitial atoms, without the formation of new phases. The complexes can lead to a considerable increase in strength due to asymmetrical lattice distortions, as well as to a consequent strong interaction with dislocations which activates intermittent flow and deformation during ageing. The latter effect is often related to an embrittlement, particularly at elevated temperatures [1]. The temperature at which strain-ageing takes place is determined by the rate of diffusion; mostly of substitutional atoms rather than interstitial ones. In this case the degree of influence of substitutional atoms is in the order: Fe, Mn, Cr, Mo, V.

In some papers [63, 64] it was proposed to divide the dislocation interaction with impurity atoms into three types:

1. Interaction resulting in the initiation of friction in the motion of dislocations.

2. Interaction resulting in the pinning (locking) of dislocations.

3. Interaction of moving dislocations with mobile impurity atoms.

In calculating the contribution made to a material's yield point, by the interaction of impurity atoms with dislocations, the first two factors play the main role.

Strengthening of the first type is due to the operation of several mechanisms:

a) strengthening related to the atomic-size misfit between the impurity and matrix, given by the parameter δ_a:

$$\delta_a = \frac{1}{a} \cdot \frac{da}{dc}, \tag{2.12}$$

where $\frac{da}{dc}$ is the lattice parameter of the solid solution as a function of the atom concentration of the dissolved element. For dilute solid solutions (c << 1), as in steels:

$$\delta_a = 2 \frac{a_m - a_r}{a_m + a_r}, \tag{2.13}$$

where a_m and a_r are the parameters for the matrix and dissolved element, respectively. On this basis, the expression for the critical flow stress of a solid solution is:

$$\tau_{cr} = 2.5G \cdot \delta_\alpha^{3/4} \cdot c, \tag{2.14}$$

b) strengthening related to the misfit of the elastic moduli of the admixture and matrix atoms, determined by the parameter:

Materials Research Forum LLC

https://doi.org/10.21741/9781644902776

$$\delta_a = \frac{1}{G} \cdot \frac{dG}{dc}, \tag{2.15}$$

For dilute solutions:

$$\delta_a = 2\frac{G_m - G_r}{G_m + G_r}, \tag{2.16}$$

where G_m and G_r are the elastic moduli of the matrix and dissolved element, respectively [65]. According to Fleischer and Hibbard [66] the strengthening due to size and elastic misfit can be calculated from:

$$\sigma_r = G \cdot \delta_s^{3/2} \cdot \frac{c^n}{m}, \tag{2.17}$$

where $m = 760$ and $\delta_s = |\delta_G| + \alpha_0|\delta_a|$ is a misfit parameter, with $\alpha_0 = 3$ for edge and $\alpha_0 = 16$ for screw dislocations while $n = 1/2$. In other work [66, 67] it is shown that the exponent n can be equal to 1; 1/2; 1/3; 0.3.

Equation (2.17) is inconvenient for calculating the solid-solution strengthening of complex alloy-steels, so an additivity of the contributions to strengthening made by separate alloying elements is therefore usually assumed by using the approximate empirical formula [1, 47]:

$$\sigma_r = \sum_{i=1}^{m}(k_i \cdot c_i), \tag{2.18}$$

where k_i is the coefficient of ferrite-strengthening reflecting the yield-point increment upon dissolving 1wt% of the i-th alloying element, and c_i is the concentration of the i-th element in the ferrite, in weight percent. The values of the coefficient k_i for various elements are determined experimentally [66, 68].

Strengthening of the second type is caused by the pinning of dislocations by impurity atoms. There are various types of pinning: elastic pinning of dislocations (Cottrell mechanism), 'chemical' pinning (Suzuki locking), electrostatic pinning, etc. [57, 69]. Cottrell elastic pinning [70, 71] plays the most substantial role for the alloys of iron.

Fig. 2.11 shows the contribution made by the presence of interstitial (carbon) and substitutional (nickel, chromium, molybdenum, vanadium) atoms as a function of the degree of deformation of quenched 38CrNi3MoV steel, as calculated using expression (2.18). It is clearly seen that, with increasing degree of deformation, the value of the contribution increases from 660 to 800MPa, due to the dissolution of 'self-tempering' cementite particles and to the entry of some of the carbon atoms into the iron lattice and their deposition on dislocations (see Chapter 1).

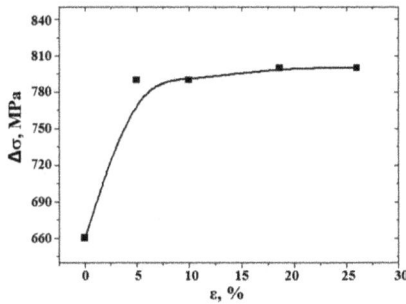

Fig. 2.11. *Contribution of solid-solution strengthening to the flow stress as a function of the degree of deformation.*

A comparison of contributions to the strain-strengthening of quenched steel is presented in Fig. 2.12. It is clearly seen that the largest contribution to strengthening is made by long-range stress fields (curve 1) and by solid-solution strengthening (curve 2). In this case the contribution of long-range stress fields increases with increasing degree of deformation; the contribution of solid-solution strengthening increases sharply in the initial stage of deformation and saturates above $\varepsilon \approx 5\%$. The contributions vary slightly with increasing degree of deformation and noticeably rank below the first two in value.

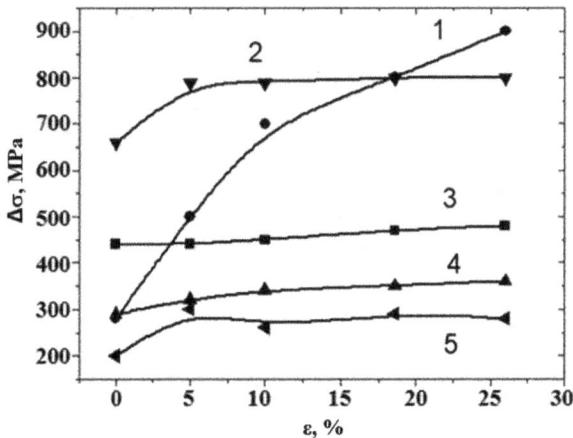

Fig. 2.12. *Contribution of long-range stress fields (1), solid-state strengthening (2), martensite crystal boundaries (3), 'forest' dislocations (4) and cementite particles (5) to the flow stress as a function of the degree of steel deformation.*

2.6. Superposition of strengthening mechanisms of quenched structural steel

As shown above, the degree of steel strain-strengthening is governed by many factors: the matrix lattice-friction, 'forest' dislocations, carbide-phase particles, intraphase boundaries, the presence of dissolved atoms of carbon and alloying elements. It is supposed that the overall yield point of the steel is a linear sum of the contributions of the separate strengthening mechanisms [1, 19, 69, 72]:

$$\sigma = \Delta\sigma_0 + \Delta\sigma_{bound} + \Delta\sigma_{disl} + \Delta\sigma_{op} + \Delta\sigma_{sol.sol.} + \Delta\sigma_{long\ field}, \qquad (2.19)$$

where $\Delta\sigma_0$ is the contribution due to the matrix lattice friction, σ_{bound} is the contribution due to intraphase boundaries, $\Delta\sigma_{disl}$ is the contribution due to the dislocation sub-structure, $\Delta\sigma_{op}$ is the contribution due to the presence of carbide phases, $\Delta\sigma_{sol.sol}$ is the contribution due to atoms of alloying elements and $\Delta\sigma_{long\ field}$ is the contribution due to long-range stress fields. As seen from equation (2.19) the additivity principle assumes independent action on the part of each of the strengthening mechanisms at the yield point. It is only a first approximation and ca lead therefore to large discrepancies between theoretical and experimental results [73].

In other papers [1, 19, 74] a quadratic summation of the contributions of strengthening mechanisms arising from equi-strong obstacles was suggested, i.e. at $\Delta\sigma_1 \approx \Delta\sigma_2$:

$$\sigma = \sqrt{(G_1^2 + G_2^2)}, \qquad (2.20)$$

Fig. 2.13 shows the strain-strengthening curves of quenched steel, as calculated theoretically (curves 1 and 2) and detected experimentally (curve 3). It is clearly seen that additive summation of the contributions of the various mechanisms of strengthening (curve 1) results in somewhat over-estimated results, the discrepancy between the dependences calculated theoretically and obtained experimentally reaching 70kg/mm^2. Application of the principle of quadratic summation of the contributions of strengthening mechanisms arising from equi-strong obstacles (curve 2, quadratic summation of the contributions arising from long-range stress-fields and solid-solution strengthening) leads to a rather qualitative and quantitative agreement between theoretical estimates and experimental data. In this case the maximum discrepancy amounts to 15kg/mm^2.

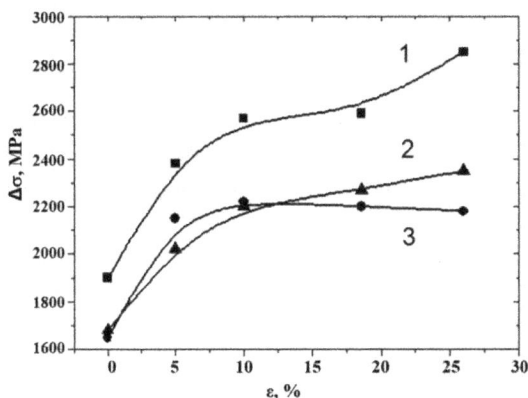

Fig. 2.13. *Curves of steel strain strengthening calculated theoretically (1,2) and determined experimentally (3); curve 1 additive summation of contributions to steel strengthening; curve 2 – quadratic summation of equi-strong contributions (inclusion long-range stress fields and solid-solution strengthening).*

Chapter 2 Conclusions

Using the results of quantitative structural analysis of a quenched steel subjected to uniaxial deformation, estimates of the effects of the strengthening mechanisms are made. Analysis of the nature of steel strain-strengthening has shown that:

1. Steel strengthening in the quenched state has a multi-factor character.

2. The largest contribution to the strain-strengthening of the steel under study is provided by a sub-structural strengthening due to long-range stress fields and solid-state strengthening due to carbon atoms.

3. The closest agreement of experimental results and estimates of strain-strengthening is observed for a quadratic summation of the contributions to strengthening made by equi-strong obstacles, the contributions arising from long-range stress fields and solid-solution strengthening.

Chapter 2 References

[1] Pickering, F.B., Physical Metallurgy and Development of Steels, Metallurgiya, 1982, 184pp.

[2] Kelly, A., Nicholson, R.B., Strengthening Methods in Crystals, Elsevier, 1971, 214pp.

[3] Fleischer, R.L., Hibbard, W.R., The Relation between the Structure and Mechanical

Properties of Metals, HMSO, 1963, 203pp.

[4] Smirnov, B.I., Dislocation Structure and Crystal Strengthening, Nauka, Leningrad, 1981, 236pp.

[5] Trefilov, V.I., Milman, Yu.V., Firstov, S.A., Physical Fundamentals of Strength of Refractory Metals, Naukova Dumka, Kiev, 1975, 315pp.

[6] Trefilov, Yu.I., Moiseev, V.I., Pechkovskii, E.P. et al. Strain Hardening and Fracture of Polycrystalline Metals, Naukova Dumka, Kiev, 1987, 248pp.

[7] Shtremel, M.A., Strength of Alloys. Part II. Deformation Handbook for Higher Schools, MISIS, 1997, 527pp.

[8] Koneva, N.A., Kozlov, E.V., Physical Nature of Stage Character of Plastic Deformation, Structural Levels of Plastic Deformation Fracture (Ed. V.E.Panin), Nauka, Siberian branch, Novosibirsk, 1990, 123-186.

[9] Gromov, V.E., Kozlov, E.V., Bazaikin, V.I., Tsellermayer, V.Ya., Ivanov, Yu.F., et al. Physics and Mechanics of Drawing and Die Stamping, Nedra, 1997, 293pp.

[10] Vladimirov, V.I., Physical Theory of Strength and Plasticity. Point Defects. Strengthening and Recovery, LPI, 1975, 120pp.

[11] Rybin, V.V., Large Plastic Deformations and Metal Fracture, Metallurgiya, 1986, 224pp.

[12] Finkel, V.M., Physical Fundamentals of Fracture Retardation, Metallurgiya, 1977, 359pp.

[13] Kaibyshev, O.V., Baliev, R.Z., Grain Boundaries and Metal Properties, Metallurgiya, 1987, 216pp.

[14] Statistical Strength and Mechanics of Steel Fracture, Transaction of Scientific Papers (Translated from German). Eds. V.Dal & V.Anton, Metallurgiya, 1986, 566pp.

[15] Hall, E.O., The deformation and ageing of mild steel: III discussion of results, Proc. Phys. Soc., 1951, 64B, 747-753. https://doi.org/10.1088/0370-1301/64/9/303

[16] Petch, N.J., The cleavage strength of polycrystals, J.Iron Steel Inst., 1953, 174, 25-28.

[17] Luke, K., Gottshteing, G., Atomic Mechanisms of Metal Plasticity, Statistical Strength and Mechanics of Steel Fracture: Transaction. (Translated from German), Eds. V.Dal & V.Anton, Metallurgiya, 1986, 14-36.

[18] Dal, V. Increase in strength at the expense of grain refinement, Statistical Strength and Mechanics of Steel Fracture: Transaction (Translated from German), Eds. V.Dal & V.Anton, Metallurgiya, 1986, 133-146.

[19] Goldshtein, M.I., Farber, B.M., Dispersion Hardening of Steel, Metallurgiya, 1079, 208pp.

Materials Research Forum LLC
https://doi.org/10.21741/9781644902776

[20] Belenkiy, B.Z., Farber, B.M., Goldshtein, M.I., Estimation of strength of low-carbon low-alloy steels by structural data, FMM, 1975, 39, 403-409.

[21] Naulor, I.R., The influence of the lath morphology on the yield strength and transition temperature on martensite-bainite steel, Met. Trans., 1979, 10A[7] 873-891. https://doi.org/10.1007/BF02658305

[22] McLean, D., Mechanical Properties of Metals, Metallurgiya, 1965, 431pp.

[23] Ashby, M.F., Mechanisms of Deformation and Fracture, Adv. Appl. Mech., 1983, 23, 118-177. https://doi.org/10.1016/S0065-2156(08)70243-6

[24] Keh, A.S., Direct Observations of Crystals, Interscience, 1962, 213pp.

[25] Bailey, J.E., Hirsch, P.B., The dislocation distribution, flow stress and stored energy in cold-worked polycrystalline silver, Phil. Mag., 1960, 53, 485-497. https://doi.org/10.1080/14786436008238300

[26] Kuhlman-Wilsdorf, D. A critical test theories of work-hardening for the case of drawn iron wire, Met. Trans., 1970, 1, 3173-3179. https://doi.org/10.1007/BF03038434

[27] Predvoditelev, A.A., The state-of-the-art of studying dislocation ensembles, Problems of Modern Crystallography, Nauka, 1975, 262-275.

[28] Lavrentev, F.F., The type of distribution as the factor determining work hardening, Mat. Sci. and Eng., 1980, 16, 191-208. https://doi.org/10.1016/0025-5416(80)90175-5

[29] Embyri, I.D., Strengthening by dislocation structure, Strengthening Method in Crystals, Applied Science Publishers, 1971, 331-402.

[30] Koneva, N.A., Kozlov, E.V., Nature of substructural strengthening, Proceedings of Higher Schools - Physics, 1982, 8, 3-14. https://doi.org/10.1007/BF00895238

[31] Koneva, N.A., Kozlov, E.V., Physics of substructural stress strengthening, Bulletin of TGASU, 1999, 1, 21-35.

[32] Kocks, U.F., Statistical treatment of penetrable obstacles, Canadian Journal of Physics, 1967, 45[2] 737-755. https://doi.org/10.1139/p67-056

[33] Strunin, B.M., Probabilistic description of internal stress field for a random arrangement of dislocations, FTT, 1971, 13[3] 923-926.

[34] Seeger, A., Mechanism of gliding and strengthening in fcc and hexagonal close-packed metals, Dislocations and Mechanical Properties of Crystals, IIL, 1960, 179-289.

[35] Hirt, J., Lotte, I., Theory of Dislocations, Atomizdat, 1972, 599pp.

[36] Ivanov, Yu.F., Kozlov, E.V., Electron-microscopic analysis of martensite phase of the 38CrNi3MoV steel, Proceedings of Higher Schools - Ferrous Metallurgy,

1991, 8, 38-41.

[37] Koneva, N.A., Lychagin, D.V., Teplyakova, L.A., Kozlov, E.V., Dislocation-disclination substructures and hardening, Theoretical and Experimental Study of Dislocations, FTI, 1984, 116-126.

[38] Vladimirov, V.I., Physical theory of strength and plasticity. Point Defects, Strengthening and Recovery, LPI, 1975, 120pp.

[39] Eshelby, J., Continuum Theory of Dislocations, TLI, 1963, 247pp.

[40] Shtremel, M.F., Hardness of Alloys - Part I: Lattice Defects, MISIS, 1999, 384pp.

[41] Hirsch, P.B., Howie, A., Nicholson, R.B. et al. Electron Microscopy of Thin Crystals, Mir, 1968, 577pp.

[42] Koneva, N.A., Kozlov, E.V., Trishkina, L.I., Lychagin, D.V., Long-range stress fields, curvature-torsion of crystal lattice and stages of plastic deformation. Methods of measurement and results, New Methods in Physics and Mechanics of Deformed Solid, Transactions of International Conference, TGU, Tomsk, 1990, 83-93.

[43] Koneva, N.A., Lychagin, D.V., Teplyakova, L.A., Kozlov, E.V., Turns of crystal lattice and stages of plastic deformation, Experimental Investigation and Theoretical Description of Dislocations, FII, 1984, 161-164.

[44] Teplyakova, L.A., Ignatenko, L.N., Kasatkina, N.F., Ivanov, Yu.F. et al. Regularities of plastic deformation of steel with a structure of tempered martensite, Plastic Deformation of Alloys. Structurally-Inhomogeneous Materials, TGU, Tomsk, 1987, 26-51.

[45] Hornbogen, E., Increase of hardness by disperse precipitates. Transaction. (Translation from German) Eds. V.Dal & V.Anton, Metallurgiya, 1986, 165-189.

[46] Orowan, E. Symposium on Internal Stresses in Metals and Alloys, Inst. Metals, London, 1948, 451pp.

[47] Tekin, E., Kelly, P.M., Tempering of Steel, Precipitation from Iron-Based Alloys, Gordon & Breach, 1965, 283pp.

[48] Ashby, M.F., Physics of Strength and Plasticity, MIT Press, Cambridge, Mass., 1969, 113pp.

[49] Eshelby, J.D., The stresses at the inclusion-matrix interface, Progress in Solid Mechanics, Vol.2, Chap.3, Wiley, New York, 1961, 534-541.

[50] Ansell, G.S., Lenel, F.V., Criteria for yielding of dispersion-strengthened alloys, Acta Met., 1960, 8[9] 612-616. https://doi.org/10.1016/0001-6160(60)90015-8

[51] Hirsch, P.B., Humphreys, F.J., Plastic deformation of two-phase alloys containing small non-deformable particles, Physics of Strength and Plasticity, Metallurgiya, 1972, 158-186.

[52] Kelly, A., Nicholson, R., Disperse Hardening, Metallurgiya, 1966, 187pp.

[53] Pyabko, P.V., Ryaboshapka, K.P., Theories of yield point of heterophase systems with coherent particles, Metallophysics, 1970, 31, 5-31.

[54] Gerold, V., Habercorn, H., On the critical resolved shear stress of solid solutions containing coherent precipitations, Phys. Status Solidi, 1966, 16[2] 675-684. https://doi.org/10.1002/pssb.19660160234

[55] Fleischer, R.L., Dislocation structure in solution hardened alloys, Electron Microscopy and Strength of Crystals. Wiley, New York, 1963, 973-989.

[56] Mott, N.F., Nabarro, F.R.N., The distribution of dislocations in slip bands, Proc. Phys. Soc., 1940, 52[1] 86-93. https://doi.org/10.1088/0959-5309/52/1/312

[57] Friedel, G., Dislocations, Mir, 1967, 643pp.

[58] Krishtal, M.A., Interaction of dislocations with impurity atoms and properties of metals, Physics and Chemistry of Material Treatment, 1975, 1, 62-71.

[59] Fleischer, R.L., Substitutional solution hardening, Acta Met., 1963, 11, 203-209. https://doi.org/10.1016/0001-6160(63)90213-X

[60] Pickering, O.F.B., Gladman, T., Iron and Steel Inst. Spec. Rep., 1963, 81, 10.

[61] Dyson, D.J., Holmes, B., J.Iron Steel Inst., 1970, 208, 469.

[62] Fleischer, R.L., Hibbard, W.R., The Relation Between the Structure and Mechanical Properties of Metals, HMSO, 1963, 261pp.

[63] Fasiska, E.J., Wagenblat, H., Dilatation of alpha-iron by carbon, Trans. Met. Soc. AIME, 1967, 239[11] 1818-1820.

[64] Kalich, D., Roberts, E.M., On the distribution of carbon in martensite, Met. Trans., 1971, 2[10] 2783-2790. https://doi.org/10.1007/BF02813252

[65] Barnard, S.J., Smith, G.D.W., Sarikaya, M., Thomas, G., Carbon atom distribution in a dual phase steel: an atom probe study, Scripta Met., 1981, 15[4] 387-392. https://doi.org/10.1016/0036-9748(81)90216-7

[66] Ridley, T., Stuart, H., Zwell, L., Lattice parameters of Fe-C austenite at room temperature, Trans. Met. Soc. AIME, 1969, 246[8] 1834-1836.

[67] Veselov, S.I., Spektor, E.Z., Dependence of austenite lattice parameter on carbon content at high temperatures, FMM, 1972, 34[5] 895-896.

[68] Vohringer, O., Macherauch, E., Struktur und Mechanische Eigenschaft von Martensite, HTM, 1977, 32[4] 153-202. https://doi.org/10.1515/htm-1977-320401

[69] Norstrom, L.A., On the yield strength of quenched low-alloy lath martensite, Scandinavian J.Met., 1976, 5[4] 159-165.

[70] Cottrell, A.H., Dislocations and Plastic Flow in Crystals, Metallurgiya, 1958, 267pp.

[71] Cottrell, A.H., Bilby, B.A. Dislocation theory of yielding and strain ageing of iron, Proc. Phys. Soc. A., 1949, 62, 49-53. https://doi.org/10.1088/0370-1298/62/1/308

[72] Prnka, T., Quantitative relations between parameters of disperse precipitates and mechanical properties of steels, Physical Metallurgy and Thermal Treatment of Steel, 1979, 7, 3-8.

[73] Toronen, T., Kotilainen, H., Nehonen, P., Combination of elementary hardening mechanisms in Fe-Cr-Mo-V-steel, Proc. Int. Conf. Martensite Trans. ICOMAT-1979, Cambridge, 1979, 2, 1437-1442. https://doi.org/10.1016/B978-1-4832-8412-5.50237-X

[74] Butler, E.R., Burke, M.G., Martensite formation at grain boundaries in sensitised 304 stainless steel, J.Physique, 1982, 43[12] 121-126. https://doi.org/10.1051/jphyscol:1982411

[75] Orowan, E., Conditions for dislocation passage of precipitates, Proc. Symp. Intern. Stress in Metals and Alloys, London, Inst. Met., 1948, 451-454.

Physics of Strain Hardening of Structural Steels
Materials Research Foundations **153** (2023)

Materials Research Forum LLC
https://doi.org/10.21741/9781644902776

Chapter 3. Evolution of phase composition and defect sub-structure of bainitic steel under deformation in uniaxial compression

3.1. Strain-hardening curves of 30Cr2Ni2MoV steel with bainitic structure

At present the fact that the strain-hardening curves of mono- and polycrystalline materials comprise several stages is widely recognized [1–7]. The question of the stages of strain-hardening curves has been studied most extensively for single crystals of fcc metals and alloys. The stages of strain-hardening have been distinguished, and the parameters characterizing them have been analyzed [8–14]. This section presents the results of studies of the strain-hardening σ–ε curve of 30Cr2Ni2MoV structural steel having a bainitic structure, and the stages of the deformation curve are analysed.

3.1.1. Strain-hardening curves for 30Cr2Ni2MoV steel with a bainitic structure

As noted in the previous chapter, the deformation of 30Kh2N2MFA structural steel with a bainitic structure was carried out via the uniaxial compression of samples having dimensions of 4 x 4 x 6mm^3 at room temperature in an Instron-type machine at a rate of 10^{-2}s^{-1}, with automatic recording of the load and sample dimensions. The results of the experiment were subjected to statistical processing, and a reliable average value of the stress was taken to be one that, with a confidence probability of 0.75, did not deviate from <σ> by more than 20MPa. When choosing the minimum number of samples required to estimate the average to a given accuracy we used Student's test for the normal distribution law.

When compressing samples, especially to high degrees of deformation, frictional forces on the end surfaces become significant. To reduce these, graphite grease was used together with gaskets made of filter paper impregnated with paraffin. Compression as a method of deformation was convenient to use since, in this case, it is possible to achieve greater deformations than when stretching. Machine strain curves, expressed in coordinates of load (P) versus the total elongation (Δl), were re-calculated and re-plotted as true stress σ versus true strain ε relationships. The strain-hardening coefficient $\theta = \frac{\partial \sigma}{\partial \varepsilon}$ was also determined.

The characteristic shape of the strain-hardening curve of 30Cr2Ni2MoV structural steel with a bainitic structure is shown in Fig. 3.1.

Mathematical processing of the strain-hardening curves shows that the σ-ε dependence is parabolic and is described by a polynomial of the fourth degree.

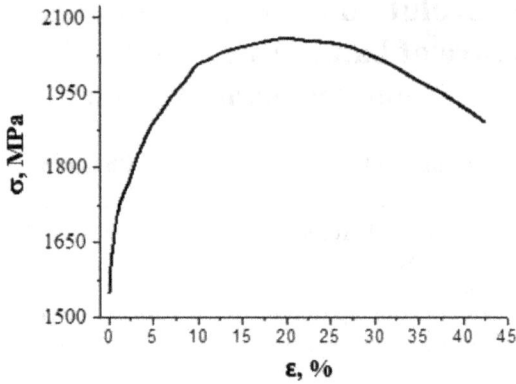

Fig. 3.1. *Characteristic strain hardening curve of
30Cr2Ni2MoV steel with a bainitic structure.*

3.1.2. Stages of plastic deformation of 30Cr2Ni2MoV steel with bainitic structure

As a rule, the strain-hardening of steel is characterized by the coefficient of strain-hardening $\theta = \frac{\partial \sigma}{\partial \varepsilon}$, which is found by differentiating the function σ–ε. Upon analyzing curves such as the one in Fig. 3.2, it is possible to distinguish two stages of strain-hardening: a stage with a parabolic σ–ε dependence (a decreasing hardening coefficient θ) and a stage with an only slightly varying, and low, value of the coefficient.

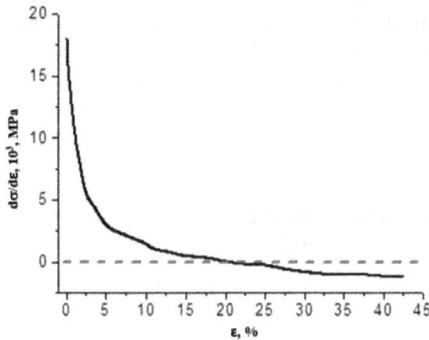

Fig. 3.2. *Dependence of the strain hardening coefficient upon the degree of deformation
of 30Cr2Ni2MoV steel with a bainitic structure.*

If one compares the form of the σ–ε and θ–ε dependences, with those observed at this stage in fcc alloys, where the stages of the flow curves have now been well studied, the above-mentioned stages should be referred to as stages III and IV. Stage III is indeed characterized by a parabolic σ–ε dependence, a rapid decrease in the hardening coefficient and a banded sub-structure. The same changes in the mechanical characteristics also occur in the steel investigated in this work (Fig. 3.2), and the structure of lower bainite, like the structure of packet-martensite, is in many respects similar to the banded sub-structure [15]. Stage IV is characterized by a constant low hardening and the development of a sub-structure with continuous and discrete misorientations, or a fragmented sub-structure. The mechanical characteristics of the steel under investigation are again analogous to those of fcc alloys. Destruction of test samples of hardened steel occurred at ε = 0.43 to 0.47, via brittle fracture at an angle of ~45° to the axis of deformation; with the formation of several large fragments.

3.2. Evolution of the structure of 30Cr2Ni2MoV steel during deformation

The kinetics of the bainitic transformation and the structures formed during this process share features of the kinetics and structures which are found for the diffusion-based pearlite and diffusionless martensitic transformations: the diffusive redistribution of carbon in austenite among its decomposition products and martensitic diffusionless transformation with the formation of a plate-like structure [16–19]. As a result of bainitic transformation under continuous cooling a multiphase structure is thus formed in the steel, comprising α-phase (a solid solution based upon the bcc crystal lattice), γ-phase (residual austenite, a solid solution based upon the fcc crystal lattice) and (in low- and medium-carbon steels) iron carbide (cementite). Like martensite, bainite is the structural basis of the high-strength state of structural steels which, under subsequent deformation, can exhibit various mechanisms of hardening and softening; creating new structural states that make it possible to vary the mechanical properties of the material over wide ranges [19–29]. Hence the need for a thorough and comprehensive analysis of the phase composition, morphology, state of the defective bainite sub-structure and their evolution during deformation.

In order to identify promising areas of application of technologies based upon the plastic deformation of steel with a bainitic structure and the choice of the most appropriate technological scheme for deformation processing for each particular material, investigations were carried out. These examined the dependence of the strain-hardening effect upon the structural state of the material before deformation and the parameters of the regime of this treatment so as to establish cause-effect relationships between the phenomena that determine the complex improvement of properties. A knowledge of the laws governing the formation of the structure and of the properties of steel during plastic deformation is in turn necessary for controlling the process of strain-hardening.

In this section we analyze (at qualitative and quantitative levels) the results obtained in the study of the phase composition and the state of the defect sub-structure of

30Cr2Ni2MoV which forms as a result of heat treatment, together with the evolution of the defect sub-structure and phase composition of this steel when subjected to plastic deformation by uniaxial compression.

As already noted above, a multiphase structure represented by the α-phase (solid solution based upon the bcc lattice), the γ-phase (residual austenite, a solid solution based upon the fcc lattice) and (in low- and medium-carbon steels) iron carbide (cementite) is formed as a result of bainite transformation during the continuous cooling of 30Cr2Ni2MoV steel [16–19]. The main phase of this class of steels is the ε-phase; the volume fraction of residual austenite varies around 10% and the volume fraction of carbide-phase particles varies from 1 to 2%. A typical image of the bainitic structure of 30Cr2Ni2MoV steel is shown in Fig. 3.3.

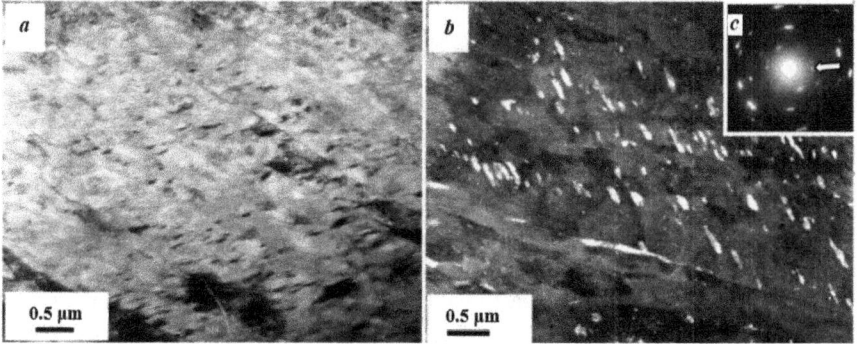

Fig. 3.3. *Electron microscopic image of the structure of 30Cr2Ni2MoV steel; a – bright field; b – dark field obtained for the [031] Fe$_3$C reflection; c – electron-microscopic diffraction pattern, the arrow indicates the reflection for which a dark field is obtained.*

The martensitic (shear) mechanism for the formation of ferrite leads to the formation of a dislocation-type sub-structure in the bainite plates, with a relatively high scalar dislocation density, amounting to ≈7·10^{10}cm^{-2} in the steel under study (Fig. 3.4a). Plastic deformation of the steel leads to an increase in the scalar dislocation density (Fig. 3.5a). The type of dislocation sub-structure does not change and the net dislocation sub-structure remains (Fig. 3.4b).

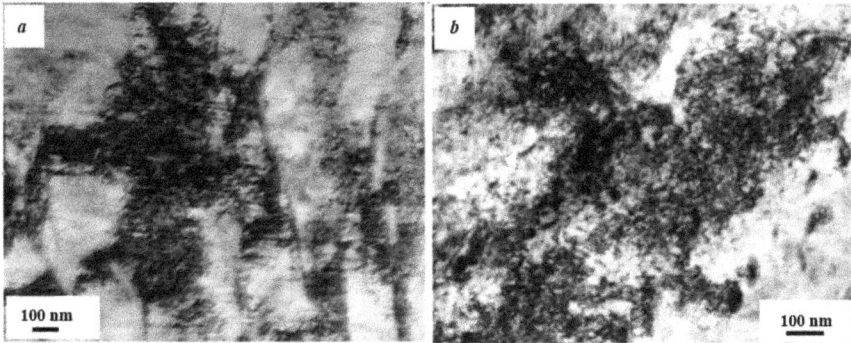

Fig. 3.4. *Electron-microscopic image of the dislocation sub-structure of bainitic steel.*

Analyzing the results presented in Fig. 3.5*a*, two sections can be distinguished with regard to the dependence of the scalar dislocation density upon the degree of deformation. In the first section (0% < ε < 18%) a linear increase in the scalar dislocation density is observed while in the second (18% < ε < 36%), equal in duration to the first section, no increase in the density of dislocations is detected. This circumstance can be attributed either to the difficulty of analyzing the dislocation sub-structure at dislocation densities greater than $\approx 10^{11}$cm^{-2}, due to the overlapping of the nuclei of closely-spaced dislocations, or to the possibility of a non-dislocation deformation mechanism.

One such mechanism can be twinning. The studies performed in this paper indeed revealed a substantial increase in the volume of material containing deformation micro-twins, with the degree of deformation exceeding $\approx 18\%$ (Fig. 3.5*b*). Typical images of the structure of steel, demonstrating the presence of micro-twins, are shown in Fig. 3.6.

The elastic stresses that occur when the $\gamma \rightarrow \alpha$ shear mechanism of the transformation occurs lead not only to the formation of a sub-structure having a high scalar dislocation density, but also to the fragmentation of bainite plates, i.e. to a partitioning of the plates into regions having a low-angle misorientation. A typical image of the fragmented structure, most clearly revealed using the methods of dark-field analysis, is shown in Fig. 3.7.

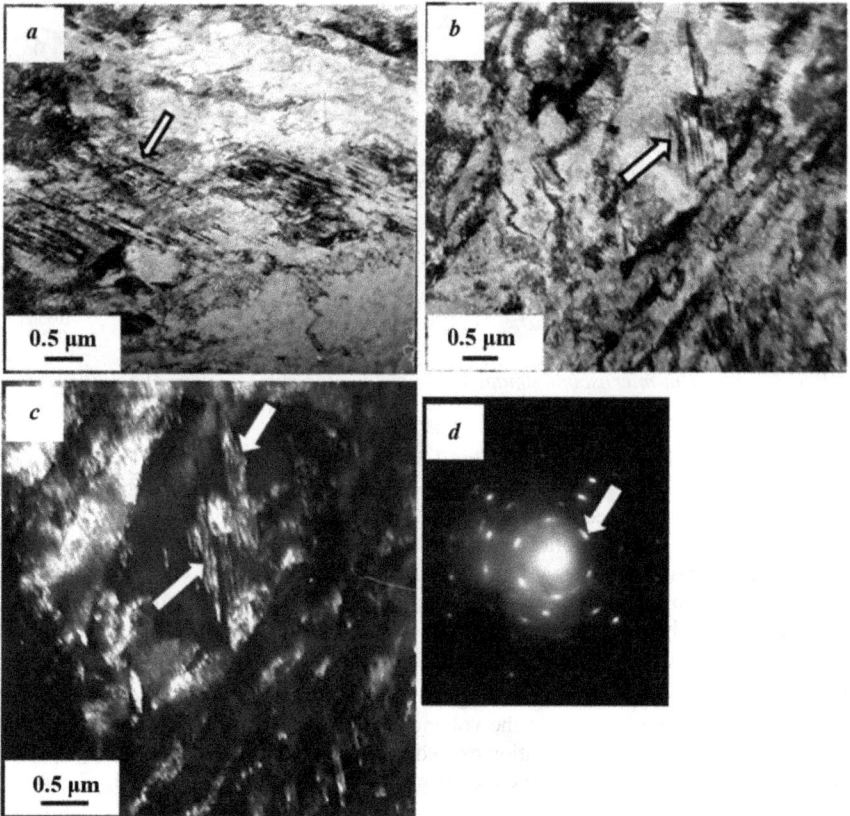

Fig. 3.6. *Electron microscopic images showing the presence of strain micro-twins in bainite plates; a, b – bright field; c – dark field produced for the [110] α-Fe reflection; d – electron diffraction micro-pattern, the arrow indicates the reflection for which the dark field is produced. The arrows in a–c indicate micro-twins within the bainite plates.*

Deformation of the steel leads to a decrease in the average longitudinal dimensions of the fragments (the transverse dimensions of the fragments are limited by the boundaries of the bainite plates and do not greatly change during deformation) (Fig. 3.8*a*). The change in the size of the fragments also reveals the existence of several stages: in the first stage, this process is characterized by a high rate while, in the second stage, it is considerably slower.

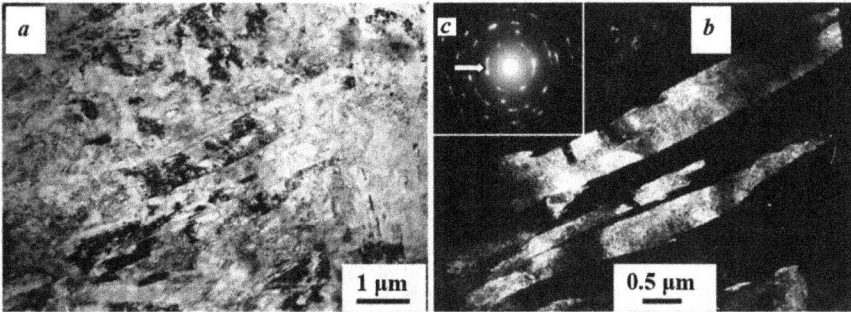

Fig. 3.7. *Electron microscopic images demonstrating the fragmentation of bainite plates; a – bright field; b – dark field obtained for the [110] α-Fe reflection; c – electron-microscopic diffraction pattern, the arrow indicates the reflection for which a dark field is obtained.*

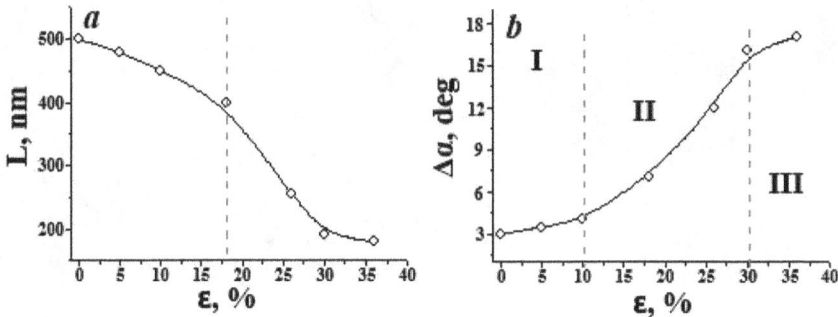

Fig. 3.8. *Dependence of the average longitudinal dimension of the fragments (a) and the magnitude of the azimuthal component of the total angle of misorientation of the bainite sub-structure (b) upon the degree of deformation.*

The change in the size of the fragments proceeds against a background of an increase in the degree of their misorientation (Fig. 3.8b). The azimuthal component of the total misorientation angle was determined from the relative size of the ε-phase refraction strands, in accordance with the procedure described in [18]. A typical image of electron-microscopic diffraction patterns, obtained by studying the structure of steel subjected to different degrees of deformation, is shown in Fig. 3.9. Analyzing the results presented in Fig. 3.8b, we can distinguish three stages in the development of this process: in stages I and III, the misorientation of the sub-structural elements increases relatively slowly, but much more rapidly in stage II.

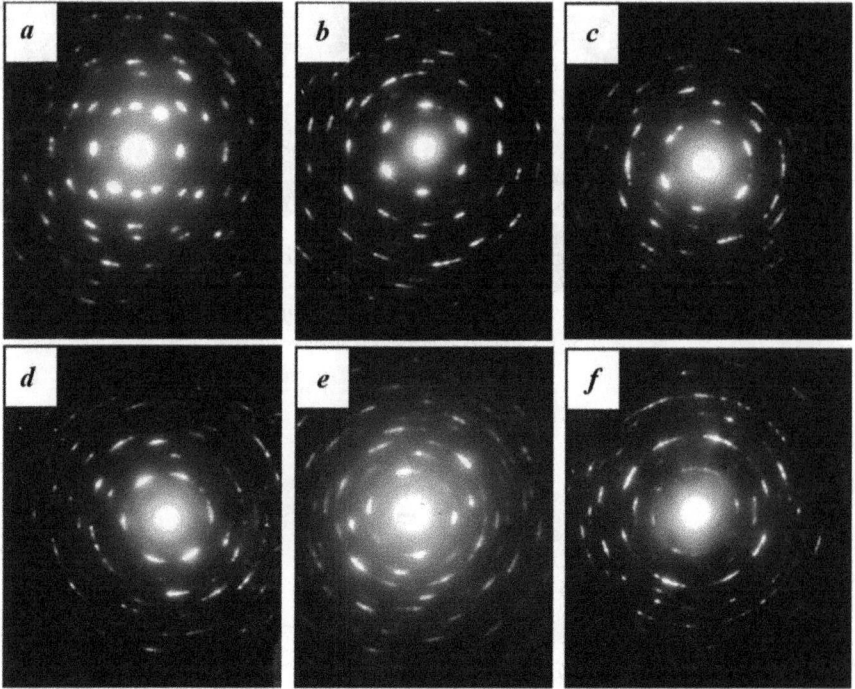

Fig. 3.9. *Electron-microscopic diffraction patterns obtained by studying the structure of steel subjected to various degrees of deformation: $a - \varepsilon = 10\%$; $b - \varepsilon = 18\%$; $c - \varepsilon = 26\%$; $d - \varepsilon = 30\%$; $e - \varepsilon = 35\%$; $f - \varepsilon = 43\%$.*

The deformation of steel is accompanied by the formation of internal stress fields, which can be revealed by the electron microscopy of thin foils and the analysis of bend extinction contours [18, 30]. Typical images of the structure of steel, showing the presence of bend extinction contours formed during plastic deformation, are shown in Figs. 3.10 and 3.11. The investigations showed that, in steel in the initial state (before deformation) and at low degrees of deformation ($\varepsilon = 5$ to 10%), the bend extinction contours are located predominantly perpendicularly (or with some deviation from this direction) to the boundaries of bainite crystals (Fig. 3.10).

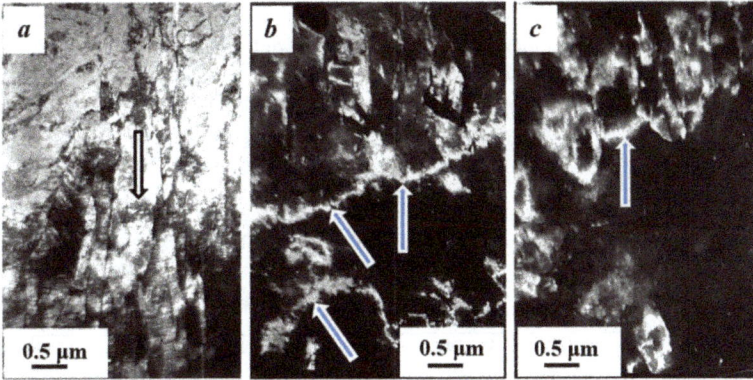

Fig. 3.10. *Electron microscopic images showing bend extinction contours; a – bright field; b, c – dark field obtained for the [110] α-Fe reflection; the bend extinction contours are indicated by arrows; ε = 10%.*

This arrangement of the bend contours indicates that the source of the internal stress fields is the interface of bainite crystals. The deformation of steel leads to a change in the morphology of the bend extinction contours. Contours located in the bulk of the bainite plates and having the form of a ring or oval (Fig. 3.11) appear.

According to the morphology of bend extinction contours it can be assumed that the deformation of steel leads to the formation of interfaces, in the bulk of the bainite crystals, which are the sources of stress fields. Such boundaries may be the interfaces of fragments, whose angle of misorientation increases with the degree of deformation of the steel, according to the results shown in Fig. 3.8*b*.

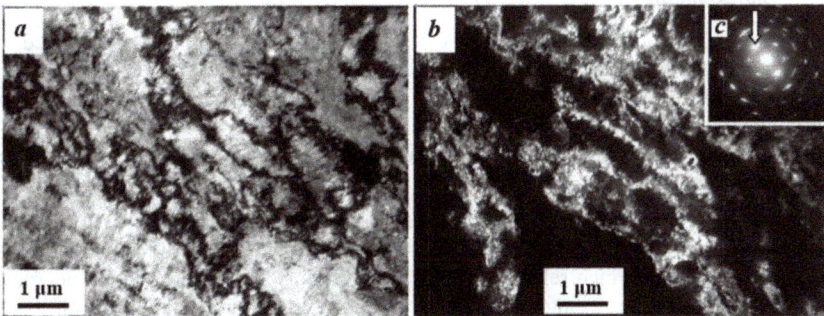

Fig. 3.11. *Electron microscopic images demonstrating bend extinction contours; a – bright field; b – dark field obtained for the [110] α-Fe reflection. The arrow (c) indicates the reflection for which the dark field was produced.*

The investigations showed that the surface-density of the contours (number of contours per unit area of the image) increases as the degree of deformation increases (Fig. 3.12*a*) and that their average transverse dimensions decrease (Fig. 3.12*b*).

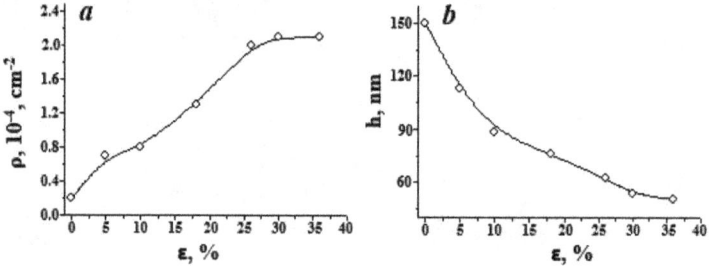

Fig. 3.12. *Dependence of the surface density of the contours (a) and their average transverse dimensions (b) upon the degree of deformation.*

The first observation indicates an increase in the number of stress concentrators in the material with increasing degree of deformation, while the second indicates an increase in the bending-torsional amplitude of the crystal lattice of the material [13, 31–34]. At the same time, as noted above, the shape of the contours and their location in the bainite plates change. If in the initial state and at low degrees of deformation the contours were located predominantly across the plates, crossing the plate from one boundary to the other, then after large deformations (18% or more) ring-shaped contours formed in the material, covering some regions within the volume of the plates.

3.2.1. Correlations and relationships in the evolution of steel structures under deformation

The deformation of steel 30Cr2Ni2MoV with a bainitic structure is accompanied, as shown above, by multiple changes in its structure and phase composition. Analysis of the structure of steel, with the determination of quantitative data (methods for determining the parameters of steel structures are given in chapter 2) made it possible to reveal a whole series of dependences and correlations. The most characteristic of them are discussed below.

Fig. 3.13 shows the dependences which characterize the deformation behaviour of a steel specimen during plastic deformation by compression. That is, the strain-hardening curve (Fig. 3.13, curve 1), which relates the increase in stress to an increase in the degree of deformation, and the change in the strain-hardening coefficient (determined by differentiating the σ–ε dependence) (Fig. 3.13, curve 2).

Materials Research Forum LLC
https://doi.org/10.21741/9781644902776

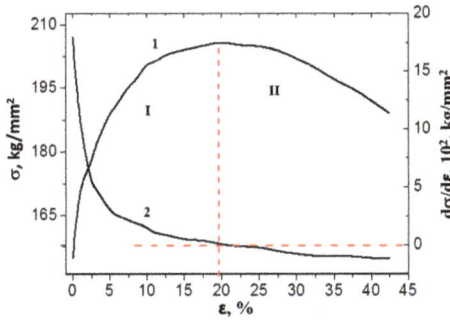

Fig. 3.13. *Strain hardening curve (1) and the change in strain hardening coefficient (2) of steel with a bainitic structure.*

Analysis of the change in the strain-hardening coefficient revealed two stages in the deformation-hardening of bainitic steel: a stage with a decreasing hardening coefficient Θ (denoted by I in Fig. 3.13, curve 2) and a stage with a slightly varying negative value of the hardening coefficient (denoted by II in Fig. 3.13, curve 2). The transition from the first stage to the second occurs at degrees of deformation ranging from 19 to 25%. It is obvious that the deformation behaviour of the samples is due to a change in the phase composition and defect sub-structure of the material. We now consider this issue in more detail.

In bainite plates, due to the shear mechanism of the $\gamma \rightarrow \alpha$ transformation, a dislocation sub-structure having a relatively high scalar dislocation density is formed which, in the investigated steel, is $\approx 7 \cdot \times 10^{10} cm^{-2}$ (Fig. 3.14, curve 1). The dislocation sub-structure consists of regular multilayer networks (Fig. 3.4*a*). The investigations showed that the type of dislocation sub-structure existing during steel deformation does not change (Fig. 3.4*b*).

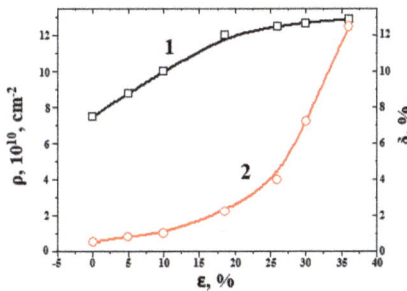

Fig. 3.14. *Dependence of the scalar dislocation density (curve 1) and the volume of material containing micro-twins (curve 2) upon the degree of deformation of steel with a bainitic structure.*

Materials Research Forum LLC

https://doi.org/10.21741/9781644902776

In spite of the very high value of the scalar density of dislocations in the steel before loading, plastic deformation of the material is accompanied by an appreciable (1.5 times) increase in this characteristic (Fig. 3.14, curve 1). The $\rho = f(\varepsilon)$ dependence is divided into two characteristic regions. Firstly the region (0% < ε < 18%) is characterized by a monotonic increase in the scalar dislocation density ranging from $\approx 7 \cdot x \ 10^{10} cm^{-2}$ to $\approx 12.0 \cdot x \ 10^{10} cm^{-2}$. Secondly the region (26% < ε < 36%) is characterized by a slightly varying scalar dislocation density, going from $\approx 12.5 \cdot x \ 10^{10} cm^{-2}$ to $\approx 12.9 \cdot x \ 10^{10} cm^{-2}$. Thirdly there is a transition region (18% < ε < 26%) with the scalar dislocation density varies from $\approx 12.0 \cdot x \ 10^{10} cm^{-2}$ to $\approx 12.5 \cdot x \ 10^{10} cm^{-2}$.

One possible cause of a significant decrease in the rate of accumulation of dislocations in the late stages of loading is the inclusion of an additional deformation mechanism. Electron microscopic micro-diffraction studies performed during this study revealed a significant increase in the volume of material containing micro-twins of deformation-origin at high deformations (Fig. 3.15).

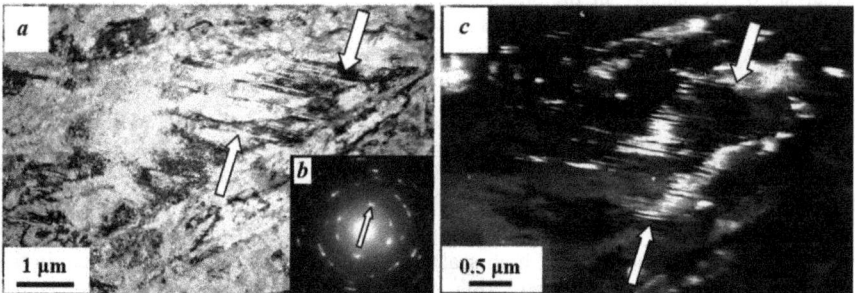

Fig. 3.15. *Electron microscopic images of the structure of steel after deformation ε = 36%; a – field; b – electron-microscopic diffraction pattern; c –dark field obtained for the [101]α-Fe reflection; the arrows indicate: (a) and (c) – micro-twins of deformation origin; (b) the reflection for which the dark field is obtained.*

Upon analyzing the results presented in Fig. 3.14, curve 2, it is noted that a sharp increase in the volume of material containing micro-twins is observed at deformation rates greater than 20 to 25%; i.e., at the stage of deformation which is characterized by a decrease in the rate of increase of scalar dislocation density.

The deformation of steel is accompanied by the fragmentation of bainite plates. An increase in the degree of deformation leads to a decrease in the average longitudinal dimensions of the fragments (the transverse dimensions of the fragments being limited by the boundaries of the bainite plates and thus essentially unchanged by deformation) Fig. 3.16, curve 1. Like the behaviour of the scalar density of dislocations, the longitudinal dimensions of the fragments change in two stages: firstly in a very substantial way (by a factor of 2.5) with the longitudinal dimensions of the fragments decreasing at deformations of up to $\varepsilon \approx 26\%$. At large values of the degree of deformation, the change in the average size of the fragments essentially ceases. It can be assumed that the sizes of the fragments reach a certain critical value (≈ 200nm). In a number of studies [35–37] it was shown that a similar structure, incapable of further evolution during deformation (the so-called critical structure), is the preferred site for the formation of micro-cracks.

The decrease in the average size of fragments proceeds with an increasing degree of their misorientation (Fig. 3.16a, curve 2). The azimuthal component of the total misorientation angle was determined by using the method described in [18]. Upon analyzing the results presented in Fig. 3.16a, it is noted that the change in the average size of the fragments, and the magnitude of the azimuthal component of the total misorientation angle of the sub-structure are related, as indicated by the results shown in Fig. 3.16, b.

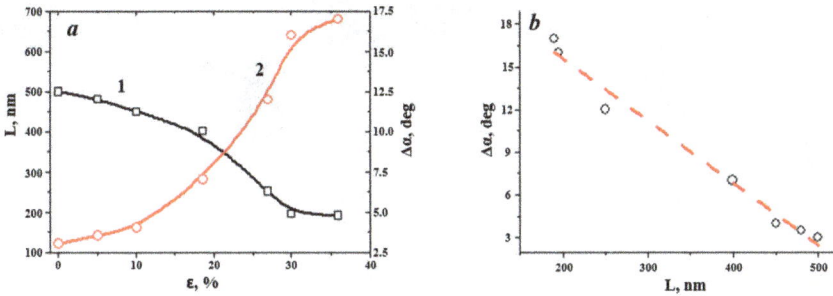

Fig. 3.16. *Dependence of the average longitudinal dimension of fragments (a, curve 1) and the magnitude of the azimuthal component of the total misorientation angle of the sub-structure (a, curve 2) upon the degree of deformation; b – linear correlation, connecting the average longitudinal dimensions of fragments and the magnitude of the azimuthal component of the total angle of misorientation of the sub-structure.*

Electron microscopic micro-diffraction analysis of steel using the thin foil method revealed bend extinction contours in the structural images. The presence of such contours indicates a bending–torsion of the lattice of the material [30–32]. It is established that in steel before deformation (initial state) and in steel after small (up to 10%) degrees of deformation, the contours are located predominantly across the plates of bainite; crossing the plate from one boundary to the other (Fig. 3.17b). Following large deformations ($\varepsilon \approx$ 18% or more) annular contours are formed in the material, covering some regions in the volume of the plates (Fig. 3.17c).

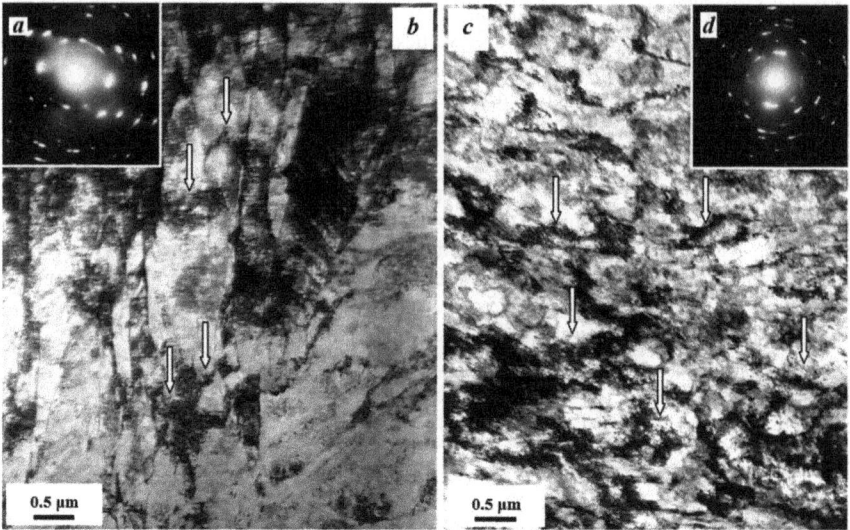

Fig. 3.17. *Electron-microscopic images showing bend extinction contours; a, d – electron diffraction micro-patterns; b, c – light fields; the extinction contours are indicated by arrows; b – ε = 10%; c – ε = 30%.*

The presence in the electron-microscopic images of the structure, of a thin foil of bend extinction contours (see Fig. 3.10, Fig. 3.11, Fig. 3.17), indicates a bending–torsion of the crystal lattice of the material; i.e., it indicates the formation, during quenching and deformation, of long-range stress fields [30, 31, 33, 38].

The bending of the crystal lattice of the material can be firstly purely elastic, due to stress fields accumulated by the incompatibility of deformation of, for example, polycrystalline grains [36, 39] or in plastic material due to non-deformable particles [40]. The sources of stress fields of elastic origin, which arise mainly in the case of inhomogeneous deformation of the material, are the joints and grain boundaries of polycrystals [41, 42], dispersed non-deformable particles [40] and, in some cases, cracks [42, 43]. There can be plastic sources if the bend is created by dislocation changes; i.e., an excess density of dislocations localized within a certain volume of material [12, 13, 15, 31–34, 38]. Thirdly there is the elastic-plastic case, when both sources of fields are present in the material.

The procedure for estimating the magnitude of long-range stress fields, along the corresponding extinction contours, is to determine the bending-torsion of the crystal lattice [33, 38]. For this purpose either the speed of movement of the extinction loop is measured, with regard to the change in the angle of inclination of the goniometer, or the width of the extinction loop. Special experiments, involving the simultaneous use of both methods, established that the width of the contour deduced from the misorientation values of quenched steels is ~1 degree. The curvature-torsion amplitude is determined by the magnitude of the continuous misorientation gradient:

$$\chi = \frac{\partial \varphi}{\partial l}, \tag{3.1}$$

where $\partial \varphi$ is the change in the orientation of the reflecting plane of the foil and ∂l is the movement of the bend contour (see Chapter 2).

If there are no dislocations in the investigated section of the foil, then there is elastic bending-torsion. In this case the amplitude of the long-range stress fields can be deduced from the formula:

$$\sigma_\tau = Gt\frac{\partial \varphi}{\partial l}, \tag{3.2}$$

where G is the shear modulus and t is the thickness of the foil.

Tests of hardened steels [15], as well as of steels subjected to various degrees and types of deformation [33, 38, 44, 45], showed that reasonable estimates of the magnitude of long-range stress fields can be made by using the approximate formula:

$$\sigma_\tau = Gt\frac{\partial \varphi}{\partial l} \approx G\frac{t}{h}, \tag{3.3}$$

where h is the transverse dimension of the bend extinction contour.

The plastic bending-torsion is provided by the local excess density of dislocations: $\rho_\pm = \rho_+ - \rho_-$. In this case the relationship [34, 38] holds:

$$\rho_\pm = \frac{1}{b}\frac{\partial \varphi}{\partial l} = \frac{\chi}{b}. \tag{3.4}$$

In the case of purely plastic bending–torsion the scalar dislocation density $<\rho>$ must be, at least, not less than the excess density determined by the formula (3.4). If the scalar dislocation density, measured locally, is smaller than the quantity $\rho_\pm$ ($\rho < \rho_+$), then elastic–plastic bending of the crystal lattice takes place. In the latter case the quantity $\rho_\pm$ is conditional, since it can never exceed $<\rho>$.

The amplitude of long-range stress fields in the case of plastic bending-torsion can be deduced from the formula [33, 38]:

$$\sigma_\tau = Gb\sqrt{\rho_\pm}. \tag{3.5}$$

The morphology of bend extinction contours thus characterizes the gradient of the bending-torsion of the crystal lattice of the material, the magnitude of the transverse dimension of the contours is the degree of bending of the crystal lattice and the amplitude of the long-range stress fields [43]. For those shown in Fig. 3.18 (curve 1) the results indicate that long-range stress fields increase over the entire steel deformation interval.

The excess dislocation density is linearly related to the curvature-torsion of the crystal lattice $\chi = b \cdot \rho_\pm$ and is proportional to the amplitude of the long-range stresses [33, 38]. The quantity χ characterizes the average amplitude of the curvature-torsion of the crystal lattice of steel. Since in the present paper the entire curvature-torsion tensor of the crystal lattice has not been measured, the number of tensor components can be judged from the density of the bend extinction contours.

A quantitative analysis of the structure of the steel made it possible to show that, with increasing deformation, the number of contours per unit area of the image increases (the surface density of the contours): Fig. 3.18*a*, curve 2. At the same time the average transverse dimensions of the contours decrease (Fig. 3.18*a*, curve 1). The first fact points to an increase in the number of stress concentrators in the material with increasing degree of deformation. The second points to an increase in the amplitude of the bending-torsion of the crystal lattice of the material [31–34]. It should be noted that both factors (the number of stress concentrators and the amplitude of the bending-torsion of the crystal lattice) change in a correlated manner (Fig. 3.18*b*).

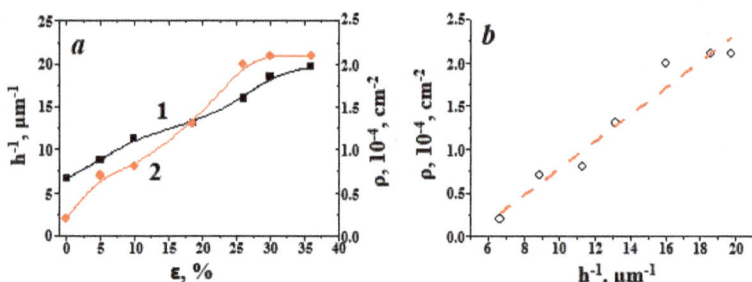

Fig. 3.18. *Dependence of the average transverse size of the contours h (a, curve 1) and of their surface density r (a, curve 2) upon the degree of deformation; b – linear correlation linking the average transverse size and the surface density of the contours.*

Analysis shows that the density of flexural extinction contours increases with increasing degree of deformation of the steel, without going to saturation in stage IV of strain-hardening. The latter indicates a constant increase in the number of components of the bending-torsion tensor that differ from zero.

The bainite transformation of steel leads to the formation of a multiphase structure [16, 18, 19]. The main component is α-phase (solid solution based upon the bcc crystal lattice) and the others are the γ-phase (solid solution based upon the fcc crystal lattice) and iron carbide (cementite).

The deformation of steel is accompanied by a significant transformation of the carbide sub-system of the material. The initially lamellar particles (with the ratio of the longitudinal dimension, L, to the transverse one, d, being $L/d \approx 8$) transform into ellipsoidal particles ($L/d \approx 5$) during the last stage of deformation. Particles having a round shape are simultaneously found within the bainite crystals (at dislocations and fragment boundaries) and their number increases with increasing degree of deformation.

The location of the cementite particles changes: the volume fraction of particles located on the boundaries of the bainite plates increases with increasing degree of deformation and, by the time of fracture of the sample, virtually all of the cementite is located on the intraphase boundaries (grain boundaries and ferrite plates) Fig. 3.19, curve 1. The increase in the volume fraction of cementite particles located at the intraphase boundaries proceeds non-monotonically: it is very quick at degrees of deformation such that $5\% < \varepsilon < 10\%$, and much slower at high degrees of deformation. The reason for an abrupt increase in the volume fraction of cementite particles at the intraphase boundaries can be transformation of the residual austenite to form the α-phase and cementite; initiated by deformation of the steel. Electron-microscopic micro-diffraction studies have indeed revealed a rapid decrease in the volume fraction of residual austenite at low ($\varepsilon \approx 10\%$) degrees of steel deformation (Fig. 3.19b).

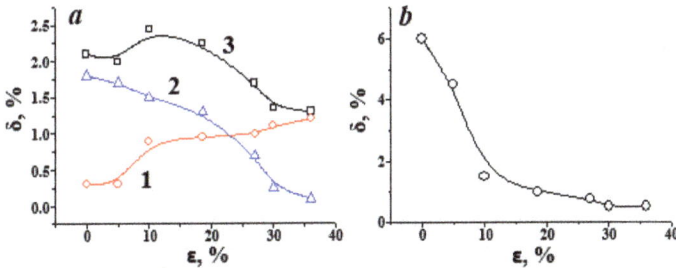

Fig. 3.19. *Effect of the degree of deformation of steel e upon the volume fraction of cementite particles (a) and interlayers of residual austenite (b); curve 1 – cementite particles located on the boundaries of ferrite crystals, curve 2 – within ferrite crystals; curve 3 shows the change in the total volume fraction of cementite in the steel.*

Attention should be paid to the behaviour of the total volume fraction of cementite (Fig. 3.19, curve 3). The initial stage of deformation ($\varepsilon \approx 10\%$) is accompanied by an increase in the total fraction of cementite particles in the steel. At large degrees of deformation the volume fraction of cementite in the steel decreases. This means that carbon in atomic form migrates to defects (dislocations, sub-boundaries, boundaries) in the crystal lattice of the steel and into a solid solution based upon the α-phase. X-ray diffraction analysis

Materials Research Foundations **153** (2023)   https://doi.org/10.21741/9781644902776

confirms this assumption. The results shown in Fig. 3.20 demonstrate an increase in the lattice parameter of the α-phase in the final stage of steel deformation.

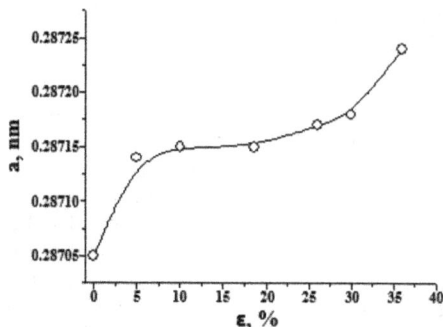

Fig. 3.20. *Dependence of the α-phase crystal lattice parameter upon the degree of deformation of steel with a bainitic structure.*

The set of facts set forth makes it possible to draw the conclusion that two competing processes occur in the steel during deformation: (1) dissolution of cementite particles formed during the bainite transformation in the volume of ferrite plates and (2) precipitation of cementite particles on the elements of the dislocation sub-structure ('strain-aging') and on the intraphase boundaries (pre-transformation of residual austenite).

Comparing the data presented in Figs. 3.13 to 3.20, obtained via quantitative analysis of the parameters of the structure of bainitic steel, it can be seen that the transition from the first stage of strain-hardening to the second stage is presaged by the following correlated change in the structural-phase state of the material. Firstly there is completion of the intensive accumulation of dislocations, and secondly there is initiation of the mechanism of deformation micro-magnetism. Thirdly there is completion of the process of fragmentation of bainite plates, and fourthly there is achievement of a maximum density of bend extinction contours. Fifthly there is a significant increase in the solid-solution hardening of the steel. Together these processes lead to the formation of regions having a critical sub-structure that is capable of forming micro-cracks, with subsequent failure of the sample.

Analyzing the results presented in this section we can therefore state that plastic deformation by uniaxial compression of 30Kh2N2MFA steel with a bainitic structure is accompanied firstly by an increase in the scalar density of dislocations and the volume of material containing deformation micro-dynamics, and secondly by a decrease in the average longitudinal sizes of the fragments and an increase in their degree of misorientation. Thirdly there is an increase in the number of stress concentrators and the bending-torsional amplitude of the crystal lattice of the material. The stages of the change in the parameters of the structure of steel are thus revealed. The assumption of a change in the mechanism of the deformation of steel is justified: in the first stage of loading (0% $< \varepsilon < 18\%$) deformation occurs mainly via the motion of dislocations while, in the second stage ($18\% < \varepsilon < 36\%$), there is the motion of dislocations ... and twinning.

Quantitative electron-microscopic micro-diffraction analysis of the evolution of the phase composition of 30Kh2N2MFA steel, taking place during plastic deformation by uniaxial compression, has been performed. It is shown that carbide transformations in the bainitic structure occur under the control of two competing processes: dissolution of cementite particles formed during the bainite transformation within the volume of ferrite plates and the release of cementite particles onto elements of the dislocation sub-structure during the 'strain-aging' process. At the same time a non-transformation of residual austenite, initiated by deformation of the steel, is observed.

3.3. Evolution of the state of the carbide phase of steel with a bainitic structure during deformation

In this section we analyze results obtained during the investigation (at qualitative and quantitative levels) of the evolution of the carbide sub-system of a medium-carbon low-alloy structural steel with a bainitic structure, subjected to plastic deformation. The parameters characterizing the deformation behaviour of its phase structure included the average size, density and volume fraction of particles of the carbide phase, the volume fraction of retained austenite (γ-phase) and the lattice parameters of the α- and γ-phases.

3.3.1. Evolution of the state of the cementite of bainitic steel during deformation

As noted above, upper and lower bainite are distinguished morphologically by their formation temperature [18, 19]. The 'lower' is accordingly bainite formed at lower temperatures, in contrast to the 'upper' bainite. A particular feature that characterizes the difference between upper and lower bainite is the location of the carbides relative to the ferrite crystals, and the shape of the carbide-phase particles. In lower bainite, part of the carbon is released from supersaturated bainitic ferrite with the formation of cementite within the ferritic (bainitic) crystals in the form of thin plates oriented at a characteristic angle (55 to 60°) to the longitudinal axis of the crystal.

In upper bainite all of the carbon is firstly displaced into the surrounding austenite phase. Carbon-enrichment stabilizes austenite. The region of carbon-enriched austenite is located between growing ferrite plates. Depending upon the composition of the alloy some of this carbon is released, during continuous cooling, as cementite or the carbon-enriched austenite is converted completely or partially into martensite. At the same time a martensitic-austenitic structural component is formed.

As a result of bainitic transformation under continuous cooling a multiphase structure is thus formed in the steel, represented by the α-phase (solid solution based upon the bcc lattice), γ-phase (solid solution based upon the fcc lattice) and (in low- and medium-carbon steels) iron carbide (cementite). A typical image of the bainitic structure of 30Kh2N2MFA steel is shown in Fig. 3.21.

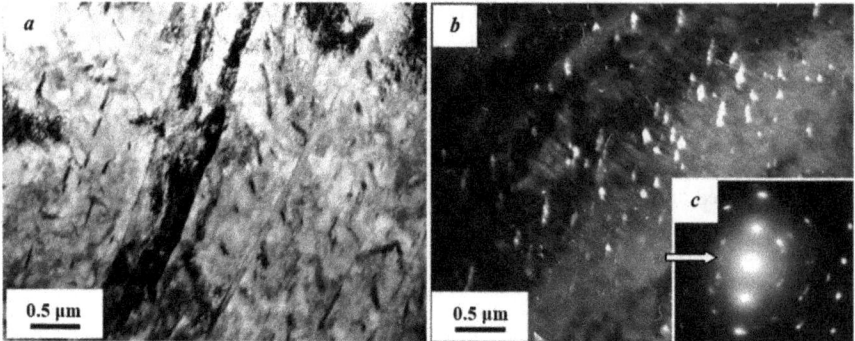

Fig. 3.21. *Electron microscopic image of the structure of 30Kh2N2MFA steel, formed as a result of cooling in air from the austenitizing temperature; a – bright field image; b – dark field image obtained for the [201]Fe₃C reflection; c – electron micro-diffraction pattern, the arrow indicates the reflection for which the dark field is obtained.*

During the deformation of steel a change in the state of the carbide phase (iron carbide, cementite) is observed. An increase in the degree of deformation leads to a decrease in the average dimensions (Fig. 3.22), density and volume fraction (Fig. 3.23) of iron carbide particles. At the same time the morphology of the particles changes. Firstly the geometrical shape of the particles is transformed: the initially lamellar particles transform into ellipsoidal ones in the last stage of deformation. Analysis of the results presented in Fig. 3.22 (inset) shows that the shape-factor $k = L/d$ decreases from $k = \approx 8$ in the initial state of the steel, to $k = \approx 5$ at sample failure.

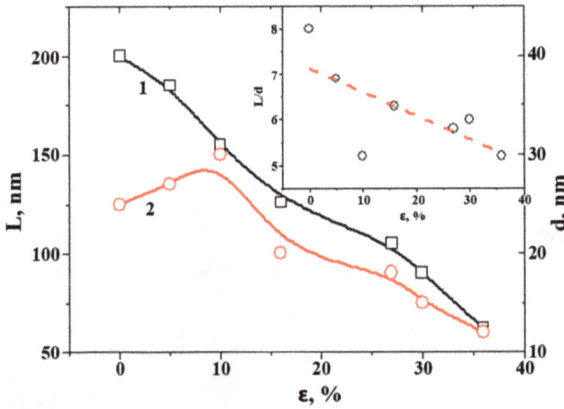

Fig. 3.22. *Dependence upon the degree of deformation of steel ε of the longitudinal L (curve 1) and transverse d (curve 2) dimensions of cementite particles located within the bulk of bainite crystals.*

Secondly the location of the cementite particles changes: with increasing degree of deformation the volume fraction of particles located on the boundaries of bainite plates noticeably increases (Fig. 3.19a, curve 1). This indicates the dissolution of cementite particles located within the volume of the plates by interaction with moving dislocations, the deposition of carbon atoms from carbide-phase particles onto dislocations (formation of Cottrell and Suzuki atmospheres) and transfer of carbon atoms by dislocations to the plate interfaces, followed by the formation of carbide-phase particles analogous to the strain-aging of steel.

The change in the total volume fraction of cementite particles (Fig. 3.19, curve 3) illustrates the complex interrelated nature of the deformation behaviour of the phase composition of steel with a bainitic structure (Fig. 3.19, curve 3). That is, an increase in the total volume fraction of cementite particles during the initial stage of deformation (ε < 10%) is associated with a deformation transformation of the residual austenite which proceeds according to: γ-Fe $\rightarrow$ α-Fe + Fe_3C. This is shown by the fact that the volume fraction of residual austenite in the steel decreases (Fig. 3.19b) and by the corresponding increase in the volume fraction of cementite particles located on the boundaries of the bainite plates (Fig. 3.19a, curve 1). At deformations of ε > 10% the volume fraction of cementite in the steel is reduced. Simultaneously any increase in the lattice parameter of the α-phase is fixed (Fig. 3.20), indicating saturation of the α-phase crystal lattice by carbon. As shown later, the carbon which is liberated upon the breakdown of cementite particles is complexly redistributed within the steel, relocating in the α-Fe lattice

Materials Research Forum LLC

https://doi.org/10.21741/9781644902776

(interstitial position) and depositing onto structural defects (Cottrell and Suzuki atmospheres, segregation at intraphase boundaries).

Thirdly, inside the bainite crystals (on dislocations and fragment boundaries), particles of round shape are found and their number increases with increasing degree of deformation (Fig. 3.23). This indicates strain-aging of the steel; i.e. the formation of new cementite particles during deformation of the material.

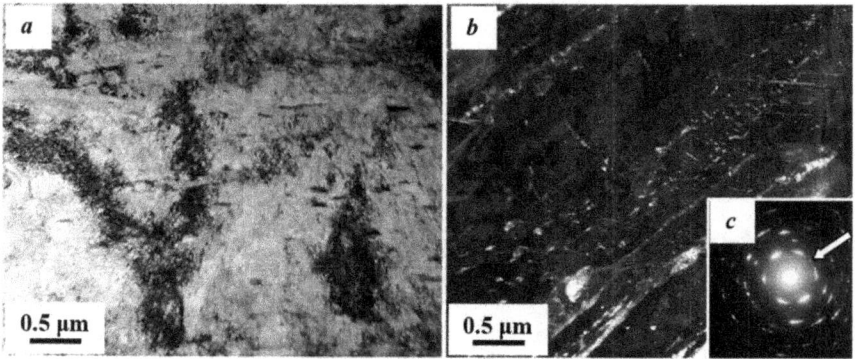

Fig. 3.23. *Electron microscopic image of the structure of 30Kh2N2MFA steel subjected to uniaxial compression at $\varepsilon \approx 36\%$; a – bright field image; b – dark field obtained for the [211] Fe_3C reflection; c is the micro-diffraction electron pattern, the arrow indicates the reflection for which the dark field is obtained.*

The foregoing observations indicate the operation of two competing processes in the steel during deformation: dissolution of cementite particles formed during the bainite transformation within the volume of the ferrite plates, and release of cementite particles onto elements of the dislocation sub-structure during the 'deformation aging' process. The total volume fraction of cementite at a given time, at high degrees of deformation ($\varepsilon > 10\%$), decreases. This means that carbon in atomic form passes to defects (dislocations, sub-boundaries and boundaries) in the lattice of the steel and to a solid solution based upon the α-phase.

As a result of quantitative electron microscopic micro-diffraction analysis of the evolution of the phase composition of 30Kh2N2MFA steel during plastic deformation by uniaxial compression it was thus shown that carbide transformation in the bainitic structure occurs under the control of two competing processes: dissolution of cementite particles formed during bainitic transformation within the volume of ferrite plates, and release of cementite particles onto elements of the dislocation sub-structure during the 'strain-aging' process. At the same time no transformation of residual austenite initiated by deformation is observed.

3.4. Redistribution of carbon during the deformation of bainitic steel

In a number of papers [19, 33, 46–55] it was shown that carbon in the steel structure can be present in solid-solutions based upon α- and γ-iron (in interstitial positions), on dislocations (in the form of Cottrell and Maxwell atmospheres), at interphase (carbide/matrix) and intraphase (grains, packets, martensite crystals) boundaries and in carbide-phase particles. The amount of carbon in solid-solutions based upon α- and γ-iron is usually estimated from the change in the lattice parameter [56–58]. Estimates of the amount of carbon in carbide particles are based upon the chemical composition of the carbide, the type of crystal lattice and the volume fraction of carbide-phase particles in the steel. For cementite (assuming a stoichiometric composition), a similar calculation was carried out in [59]. An estimate of the amount of carbon located at defects (dislocations and interfaces) is the most difficult and essentially does not enjoy a direct experimental definition. This difficulty is resolved by using indirect methods (e.g. internal friction, and micro-X-ray spectral analysis) [46, 48–51, 55] and also by making theoretical estimations. The most complete analysis of the redistribution of carbon in unalloyed steels as a function of the tempering temperature was carried out in [48] for the case of alloyed steels (quenched and low-temperature tempered) in [60, 61]. In [62] the results of quantitative studies of the structural-phase state of hardened 38KhN3MFA steel are presented; the locations of carbon are identified and its redistribution is analyzed, depending upon the austenitization temperature. In [33] similar estimates were made for hardened steel, subjected to various degrees of plastic deformation by uniaxial compression.

As already noted, as a result of bainitic transformation a multiphase structure is formed in the steel and its the main phases are the α-phase (solid solution based upon the bcc crystal lattice), the γ-phase (residual austenite, a solid solution based upon the fcc crystal lattice) and iron carbide (in low- and medium-carbon steels it is cementite) [16–19, 63]. A typical image of the bainitic structure formed in 30Kh2N2MFA steel is shown in Fig. 3.24.

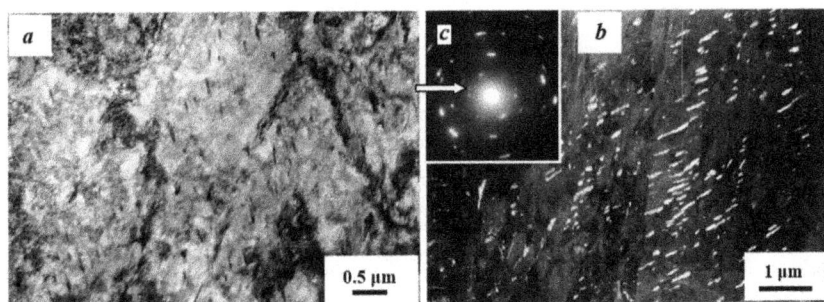

Fig. 3.24. *Electron microscopic image of steel 30Kh2N2MFA structure; a – bright field image; b – dark field obtained for the [031] Fe_3C reflection; c is the micro-diffraction electron pattern, the arrow indicates the reflection for which the dark field is obtained.*

The deformation of steel, as shown in previous sections, is accompanied by the following processes. Firstly there is an increase in the scalar density of dislocations from $7.0 \cdot \text{x} \ 10^{10} \text{cm}^{-2}$ to $12.9 \cdot \text{x} \ 10^{10} \text{cm}^{-2}$. The type of dislocation sub-structure (dislocation networks) does not change. Secondly there is a fragmentation of bainite plates. An increase in the degree of deformation leads to a decrease in the average longitudinal dimensions of fragments, from 450nm to 200nm, and an increase in the extent of their misorientation from 3° to 17°. Thirdly there is a non-transformation of residual austenite, with the formation of α-phase and cementite particles. Fourthly there is a decrease in the volume fraction of cementite particles located in bainite plates, from 1.8% to 0.1%. Fifthly there is an increase in the volume fraction of cementite particles, located at the boundaries of bainite plates, from 0.3% to 1.2%. Finally there is an increase in the α-phase lattice parameter.

The revealed quantitative patterns of change in the parameters of the structure of steel during plastic deformation made it possible to carry out studies aimed at analyzing the distribution of carbon atoms within the structure of deformed steel.

Estimations of the relative contents of carbon atoms in the structural elements of steel were carried out on the basis of the expressions summarized in Table 3.1. The results of the estimations are shown in Fig. 3.25.

***Table 3.1.** Method of analysis of the distribution of carbon in steel*

Region with carbon	Assessment equation	References
Solid solution based upon α-iron	$\Delta C_\alpha = \Delta V_\alpha \dfrac{a_\alpha - a_\alpha^0}{39 \pm 4} \cdot 10^3$ *	[55, 56, 33]
Solid solution based upon γ-iron	$\Delta C_\gamma = \Delta V_\gamma \dfrac{a_\gamma - a_\gamma^0}{44} \cdot 10^3$	[57, 58]
Carbide particles	$\Delta C_\kappa = \Delta V_i \cdot k_i;$ $k(Fe_3C) = 0.07$	[33, 59, 62]
Elements of defect structure	$\Delta C_d = C_0 - (\Delta C_\alpha + \Delta C_\gamma + \Delta C_k)$	[33, 62]

*Here ΔV_α, ΔV_γ, ΔV_i are the volume fractions of the α-Fe, γ-Fe and carbide phases, respectively; a_α and a_γ are the actual lattice parameters of the α- and γ-phases, respectively; $a_\alpha^0 = 0.28668 \ nm$; $a_\gamma^0 = 0.3555 \ nm$; C_0 is the average carbon content of the steel.

These evaluations showed that, with increasing deformation, the number of carbon atoms located in a solid-solution based on α-iron (Fig. 3.25, curve 1) forming cementite particles located on intra-phase boundaries (Fig. 3.25, curve 2) and located on defects of the crystal structure (Fig. 3.25, curve 3) increases. The number of carbon atoms forming cementite particles lying in the bulk of bainite plates (Fig. 3.25, curve 4) and located in a solid-solution based upon γ-iron (Fig. 3.25, curve 5), decreases. The plastic deformation of steel with a bainitic structure is consequently accompanied by a significant redistribution of carbon atoms. If in the initial state the major quantity of carbon atoms was concentrated in cementite particles, then in the final stage of deformation the preferred location of carbon was the α-iron-based crystal lattice.

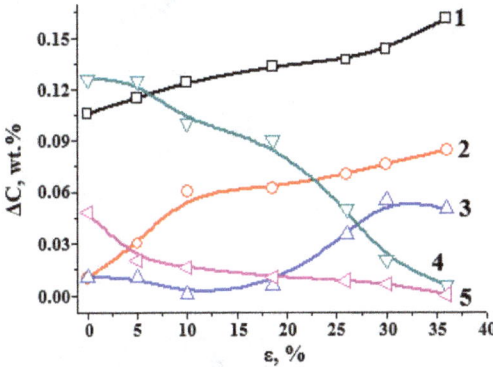

Fig. 3.25. *Effect of the degree of deformation upon the concentration of carbon atoms in a crystal lattice based upon α-Fe (1), in cementite particles lying on intraphase boundaries (2), in structural defects (3), in cementite particles within the bulk of bainite plates (4) and in a crystal lattice based upon γ-Fe (5).*

These results of a quantitative analysis of structural parameters, revealed by study of the structure and phase composition of steel with a bainitic structure subjected to plastic deformation, thus allowed us to monitor the redistribution of carbon atoms in the structure of the steel during plastic deformation. It has been established that, with increasing degree of deformation, the number of carbon atoms located in a solid solution based upon α-iron forming cementite particles located on the internal phase boundaries and the number of carbon atoms located on defects of the crystal structure increases. Meanwhile the number of carbon atoms forming cementite particles lying within the bulk of bainite plates, and located in a solid-solution based upon γ-iron, decreases.

The redistribution of carbon atoms during the deformation of bainitic steel, with migration to defects in the crystal structure, was identified by studying deformed samples subjected to subsequent annealing at room temperature for ≈15 years. The samples fractured under deformation (ε = 0.26%) following the tempering contained carbide-phase particles which were located along the dislocation lines (Fig. 3.26).

Estimations based upon electron-microscopic images of the structure of steel showed that the particle sizes are 2.0 to 2.5nm. Micro-diffraction electron patterns obtained from these precipitate data show diffusion strands located at the site of carbide-phase reflections (Fig. 3.26b). This may indicate firstly 'floating' parameters of the crystal lattice of the precipitated particles, indicating a variable elemental composition of the particles. Secondly it may be related to the defect structure of the particles or, thirdly, to their small dimensions.

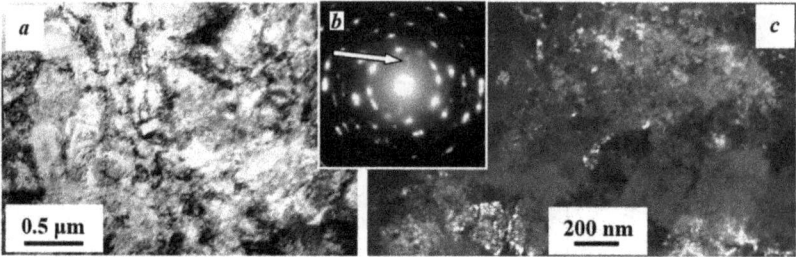

Fig. 3.26. *Structure formed in bainitic steel subjected to plastic deformation (ε = 26%) and subsequently held at room temperature for 15 years; a – bright field; b – electron-microscopic diffraction pattern; c is the dark field obtained for the [110] Fe$_3$C reflection. In (b) the arrow indicates the reflection for which the dark field is obtained.*

As the degree of deformation increases the rate of redistribution of carbon atoms increases, which is reflected by the structure of steel subjected to plastic deformation (ε = 36%) and subsequently held at room temperature for 15 years (Fig. 3.27). Particles of the carbide phase increase in size and are located predominantly at intraphase interfaces (grain boundaries, bainite plates, fragments).

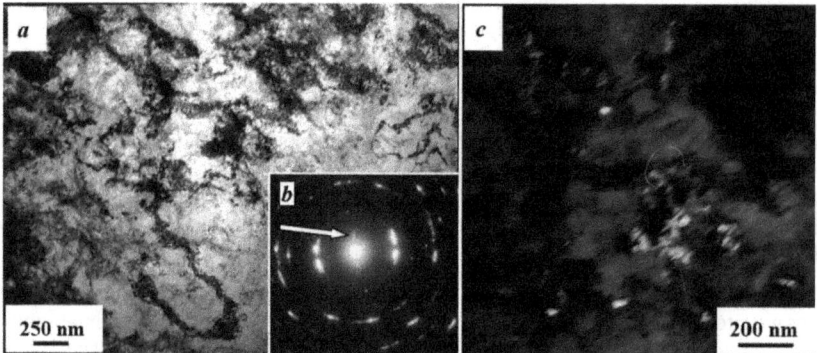

Fig. 3.27. *Structure formed in bainitic steel subjected to plastic deformation (ε = 36%) and subsequently held at room temperature for 15 years; a – bright field; b – electron-microscopic diffraction pattern; c is the dark field obtained for the [110] Fe$_3$C reflection. In (b) the arrow indicates the reflection for x the dark field is obtained.*

The process of phase separation of the material is enhanced with increasing degree of deformation. The preferential arrangement of cementite particles in steel samples subjected to plastic deformation at $\varepsilon = 43\%$ and subsequently held at room temperature for 15 years is at the boundaries and joints of grain boundaries and fragments. The size of the regions of the material having the predominant arrangement of cementite particles is 100 to 150nm, and the regions are distributed randomly throughout the bulk of the material (Fig. 3.28).

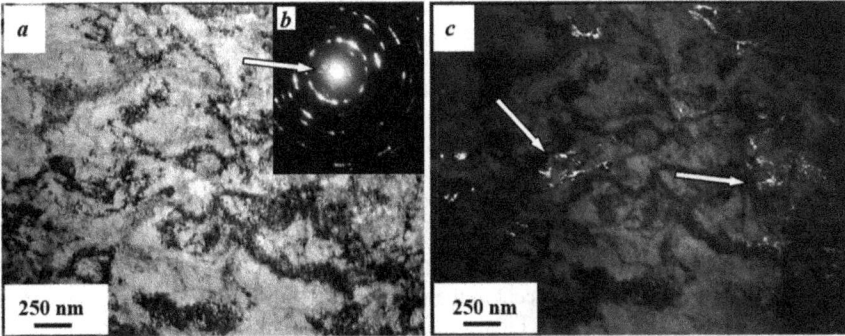

Fig. 3.28. *Structure formed in bainitic steel subjected to plastic deformation ($\varepsilon = 43\%$) and subsequently held at room temperature for 15 years; a – bright field; b – electron-microscopic diffraction pattern; c is the dark field obtained for the [110] Fe_3C reflection. In (b) the arrow indicates the reflection for which the dark field is obtained.*

3.5. Localization of the plastic deformation of bainitic steel

It was shown [44, 45] that, under conditions of intense plastic deformation by drawing at room temperature, a characteristic response of 08G2S (ferritic-pearlitic) steel is the formation of elongated regions having an ultra-dispersed structure; deformation channels. Such regions in the dark-field images of the matrix reflections appear as a speckled contrast. Micro-diffraction electron patterns obtained from such regions have, as a rule, a quasi-ring structure. The deformation channels are locations of deformation-localization. Such regions extend to a few tens of micrometres in length and have a diameter of up to 1µm. With increasing degree of deformation, the average dimensions of the deformation channels increase [44, 45]. In the deformation channel the sub-structure is also fragmented, but the dimensions of the fragments are much smaller than those in the bulk material. In addition the fragments in the deformation channel are isotropic in shape. Judging by the size of the fragments it can be assumed that the shift in the strain channel is several tens of times greater than the average. The difference in the shape of the fragments in the matrix (highly anisotropic) and channels (isotropic) indicates differing mechanisms of formation. The isotropy of the shapes of the fragments in the channel makes it possible to presume other temperature conditions for their formation. If the anisotropic fragments are the result of cold deformation, the isotropic fragments are the result of warm deformation.

The following peculiarity of the structure of the deformation channels is related to the behavior of the bend extinction contours within them. It is noted that bend extinction contours indicate regions having the same orientation of specific reflection planes with respect to the incident electron beam [18, 30]. It has been established [44, 45] that, in both the deformation channel and in the regions adjacent to it, there are segments of one orientation or close to it, which are elongated approximately parallel to the long side of the channel. From the viewpoint of hydrodynamics, such areas are similar to current lines in laminar flow. Because drawing tends to generate a significant number of areas with a turbulent flow, this comparison sheds light on the nature of the deformation channels. That is, the deformation conditions within them are such that the deformation-work is lower there than it is in neighbouring sections. It is assumed that the main role here is played by local heating of the material [44, 45].

Another feature of the deformation channels is significant stress fields localized within them and within the regions adjacent to them. Two relaxation mechanisms of these stress fields have been noted [44, 45]. Firstly fragmentation, in which chains of fragments of small dimension and similar orientation are formed along the strain channel and, secondly, by the propagation of micro-cracks.

Similar structural formations (deformation channels) were also found during the study of hardened 38KhN3MFA, deformed using uniaxial compression [33]. An example of an electron-microscopic image of the deformation channel of hardened 38KhN3MFA steel is shown in Fig. 3.29.

Materials Research Forum LLC
https://doi.org/10.21741/9781644902776

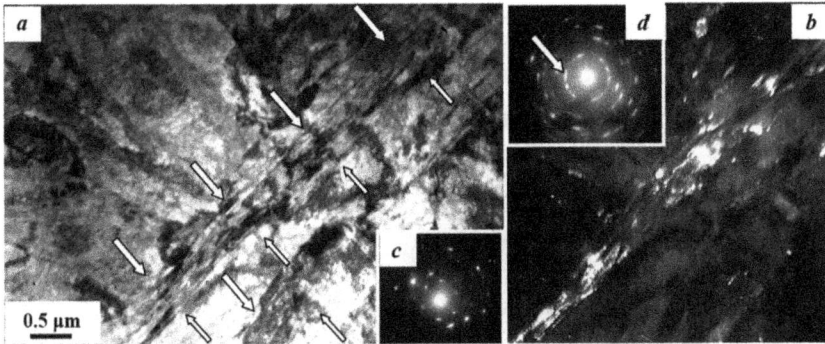

Fig. 3.29. *Deformation channels formed in hardened 38KhN3MFA steel; ε = 18.6%; a –
bright field; b – dark field obtained for the [110] α-Fe reflection; c, d – electron-
microscopic diffraction patterns. In (a) the arrows indicate the channels of deformation;
in (d) the arrow indicates the reflection for which a dark field is obtained. The electron-
microscopic diffraction pattern (c) is obtained from a foil region far from the
deformation channel; the electron-microscopic diffraction pattern (d) is from the location
of the deformation channel.*

As shown in [33], deformation channels which were revealed during the study of
hardened 38KhN3MFA steel, deformed by uniaxial compression, have the form of
elongated regions having transverse dimensions of ~0.5μm. The deformation channel has
a layered structure, recalling that of a martensite packet. The layers are formed of
crystallites whose dimensions range from 50 to 100nm. The annular structure of the
electron micro-diffraction pattern obtained from the localization region of the
deformation channel (Fig. 3.29d) indicates a predominantly high-angle misorientation of
the crystallites forming it. It is important to note that, in the areas of material adjacent to
the deformation channel, the steel structure is similar in morphology to that of the initial
state, i.e. crystals of packet and plate martensite are detected. The electron micro-
diffraction pattern obtained from the foil region adjacent to the channel is a point-like
one, characteristic of a polycrystalline material (Fig. 3.29c). With increasing degree of
deformation the volume of material occupied by deformation channels increases,
reaching a few tens of percent at the moment of fracture of the steel.

The formation of similar structures, called deformation channels in [33, 44, 45], is also
observed during the deformation of 30Kh2N2MFA steel with a bainitic structure (Fig.
3.30).

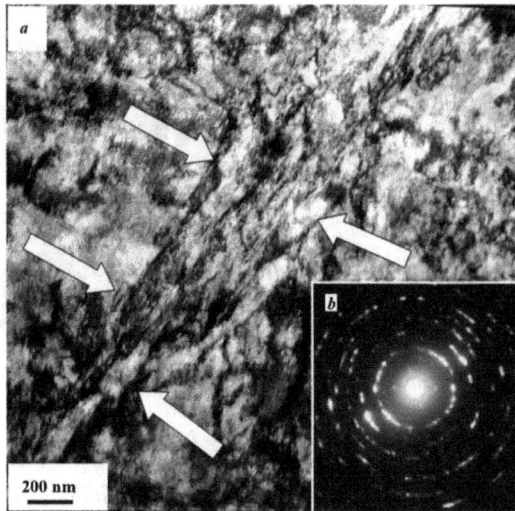

Fig. 3.30. *Deformation channels formed in bainitic 30Kh2N2MFA steel; ε = 43%; a – bright field; b – electron-microscopic diffraction pattern. In (a) the arrows indicate the deformation channels.*

Deformation channels are formed, in a given steel, at degrees of deformation degrees of 36% or more. The structure of the deformation channels of bainitic 30Kh2N2MFA steel is similar to the structure of the channels observed during the deformation by drawing of 08G2S steel in the ferritic–pearlitic state [44, 45] and in hardened 38KhN3MFA steel at room temperature [33], thus indicating the common nature of their formation.

Chapter 3 Conclusions

Analysis of the results presented in this chapter permits one to draw the following conclusions:

1. Studies of 30Kh2N2MFA steel with a bainitic structure, subjected to plastic deformation by uniaxial compression, which were performed using electron-diffraction microscopy of thin foils and X-ray diffraction analysis revealed the complex interrelated nature of the evolution of the phase composition and the defect sub-structure of the material. This manifests itself at the macro (whole sample), meso (bainite plates, interlayer residual austenite), micro (defect sub-structure of bainite plates, carbide-phase particles and nano (redistribution of carbon atoms during the fracture of particles of carbide) structural levels;

Materials Research Forum LLC
https://doi.org/10.21741/9781644902776

2. Stages in the changes of the parameters of the steel structure are revealed. It is shown that the transition from the first stage of strain-hardening to the second is presaged by the following correlated changes in the structural state of the material: firstly the completion of an intensive accumulation of dislocations, secondly the initiation of the mechanism of deformation micro-twinning, thirdly the fragmentation of bainite plates, fourthly the achievement of a maximum density of bend extinction contours and fifthly by a significant increase in the solid-solution hardening. These processes together lead to the formation of regions having a critical sub-structure that is capable of forming micro-cracks, with subsequent fracture of the sample.

3. Quantitative electron-microscopic micro-diffraction analysis of the evolution of the phase composition of 30Kh2N2MFA steel during plastic deformation by uniaxial compression has been performed. It is found that carbide transformations within the bainitic structure occur under the control of two competing processes: dissolution of cementite particles formed during bainite transformation within the bulk of ferrite plates and the release of cementite particles to elements of the dislocation sub-structure during the 'deformation-aging' process. At the same time a non-transformation of residual austenite, initiated by deformation, is observed.

4. It has been established that, with increasing degree of deformation, the longitudinal dimensions of fragments of bainite plates are reduced: increasing the density of micro-twins, the scalar and excess densities of dislocations, the linear density of flexural extinction contours and the amplitude of internal long-range stress fields;

5. The formation of channels of localized deformation, special states of material located along the interfaces of adjacent bainite plates or grain boundaries, has been revealed during the deformation of steel.

6. It is shown that prolonged holding at room temperature of samples of deformed steel is accompanied by a stratification of the material along the carbon, with the formation of regions having a preferential arrangement of carbide-phase particles.

Chapter 3 References

[1] Sachs, G., Weerts, J., Die Verfestigungskurven. Kupfer, Silber, Gold, Z.Physik, 1930, 62, 473-481. https://doi.org/10.1007/BF01339674

[2] Stepanov, A.V., Die plastischen Eigenschaften der Silberchlorid- und Natriumchlorid-Einkristalle, Phys.Z. Sowjetunion, 1935, 8[1] 25-40.

[3] Seeger, A., The mechanism of sliding and hardening in cubic face-centered and hexagonal close-packed metals, in: Dislocations and Mechanical Properties of Crystals, IIL, Moscow, 1960, 179-289.

[4] Jaoul, B., Gonzalez, D., Deformation plastique de monocristaux de fer, J.Mech. Phys. Sol., 1961, 9, 16-38. https://doi.org/10.1016/0022-5096(61)90036-9

[5] McLean, D. Mechanical Properties of Metals, Metallurgiya, 1965, 431pp.

Materials Research Forum LLC
https://doi.org/10.21741/9781644902776

[6] Ivanova, V.S., Ermishkin, V.A., Strength and Plasticity of Refractory Metal Single Crystals, Metallurgiya, 1975, 80pp.

[7] Pavlov, V.A., Physical Basis of Cold Deformation of BCC Metals, Nauka, 1978, 208pp.

[8] Vasileva, A.G., Deformation Hardening of Hardened Structural Steels, Mashinostroenie, 1981, 231pp.

[9] Bell, G.F., Experimental Foundations of Mechanics of Deformable Solids: Part II, Nauka, 1984, 431pp.

[10] Koneva, N.A., Lychagin, D.V., Zhukovsky, S.P., Kozlov, E.V., Evolution of dislocation structure and stages of plastic flow of polycrystalline iron-nickel alloy, Physics of Metals and Metal Science, 1985, 60[1] 171-179.

[11] Trefilov, V.I., et al. Deformation Hardening and Destruction of Polycrystalline Metals, Naukova Dumka, 1987, 248pp.

[12] Koneva, N.A., Kozlov, E.V., Physical nature of phase character of plastic deformation, Structural levels of plastic deformation and fracture, (Ed. V.E.Panin), Nauka, Siberian branch, 1990, 123-186.

[13] Koneva, N.A., Kozlov, E.V., Physics of substructural hardening, Bulletin TGASU, 1999, 1, 21-35.

[14] Berner, R., Kronmüller, G., Plastic Deformation of Single Crystals, Mir, Moscow, 1969, 272pp.

[15] Kozlov, E.V., Popova, N.A., Ivanov, Yu.F., Teplyakova, L.A., Band substructure and lath martensite structure. Comparison of evolution ways, Proceedings of Higher Schools - Physics, 1992, 10, 13-19. https://doi.org/10.1007/BF00559881

[16] Bhadeshia, H.K.D.H., Bainite in Steels. 2nd ed. The Institute of Materials, London, 2001, 460pp.

[17] Schastlivtsev, V.M., et al., Residual Austenite in Alloy Steels, Ekaterinburg, UrB RAS, 2014, 236pp.

[18] Utevsky, L.M., Diffraction Electron Microscopy in Metal Science, Metallurgiya, 1973, 584pp.

[19] Kurdyumov, V.G., Utevskii, L.M., Entin, R.I., Transformations in Iron and Steel, Nauka, 1977, 236pp.

[20] Houdremont, E., Special Steels. Vols. I and II, (Russian translation), Metallurgiya, 1966, 1274pp.

[21] Meskin, V.S., Fundamentals of Alloying Steel, Metallurgiya, 1964, 684pp.

[22] Lysak, L.I., Nikolin, B.I., Physical Basis of Heat Treatment of Steel, Tekhnika, Kiev, 1975, 304pp.

[23] Petrov, Yu.N., Defects and Diffusionless Transformation in Steel, Naukova Dumka, 1978, 267pp.

[24] Blanter, M.E., Phase Transformations during Heat Treatment of Steel, Metallurgiya, 1962, 268pp.

[25] Novikov, I.I., Theory of Heat Treatment of Metals, Metallurgiya, 1978, 392pp.

[26] Lakhtin, V.M., Metallurgy and Heat Treatment of Metals, Metallurgiya, 1977, 407pp.

[27] Gulyaev, A.P., Metal Science, Metallurgiya, 1978, 647pp.

[28] Pickering, F.B., Physical Metallurgy and Processing of Steels, Metallurgiya, 1982, 184pp.

[29] Schastlivtsev, V.M., Mirzaev, D.A., Yakovleva, I.L., Structure of Heat-Treated Steel, Metallurgiya, 1994, 288pp.

[30] Hirsch, P., Howie, A., Nicholson, R. et al. Electron Microscopy of Thin Crystals, Mir, Moscow, 1968, 574pp.

[31] Koneva, N.A., Kozlov, E.V., Nature of substructural hardening, Proceedings of Higher Schools - Physics, 1982, 8, 3-14. https://doi.org/10.1007/BF00895238

[32] Koneva, N.A., Kozlov, E.V., Trishkina, L.I., Lychagin, D.V., Long-range stress fields, curvature-torsion of crystal lattice and stages of plastic deformation. Methods of measurement and results, New Methods in Physics and Mechanics of Deformed Solid, Transactions of international conference, TGU, Tomsk, 1990, 83-93.

[33] Ivanov, Yu.F., Kornet, E.V., Kozlov, E.V., Gromov, V.E., Hardened Structural Steel: Structure and Strengthening Mechanisms, SibGIU, Novokuznetsk Publishing house, 2010, 174pp.

[34] Koneva, N.A., Kozlov, E.V., Trishkina, L.I., Lychagin, D.V., Long-range stress fields, curvature-torsion of crystal lattice and stages of plastic deformation. Methods of measurement and results, New Methods in Physics and Mechanics of Deformed Solid. Transactions of international conference, TGU, Tomsk, 1990, 83-93.

[35] Terentev, V.F., Model of the Physical Fatigue Limit of Metals and Alloys, Dokl. AN SSSR, 1969, 185[2] 324-326.

[36] Rybin, V.V., Large Plastic Deformation and Fracture of Metals, Metallurgiya, 1986, 224pp.

[37] Rybin, V.V., Vergazov, A.N., Likhachev, V.A., Ductile fracture of molybdenum as a result of structure fragmentation, Fiz. Met. Metalloved., 1974, 37[3] 620-624.

[38] Gromov, V.E., Kozlov, E.V., Bazaikin, V.I., Tsellermayer, V.Ya., Ivanov, Yu.F. et al. Physics and Mechanics of Drawing and Die Stamping, Nedra, 1997, 293pp.

[39] Panin, V.E., Likhachev, V.A., Grinyaev, Yu.V., Structural Levels of Deformation of Solids, Nauka, Novosibirsk, 1985, 229pp.

[40] Eshelby, J., The Continuum Theory of Dislocations, ILI, Moscow, 1963, 247pp.

[41] Vladimirov, V.I., The Physical Theory of Strength and Plasticity. Point Defects, Hardening and Recovery, LPI, Leningrad, 1975, 120pp.

[42] Shtremel, M.A., Strength of Alloys: Part I. Lattice Defects, MISIS, Moscow, 1999, 384pp.

[43] Finkel, V.M., Physical Basis of Inhibition of Fracture, Metallurgiya, 1977, 359pp.

[44] Gromov, V.E., Kozlov, E.V., Panin, V.E., Ivanov, Yu.F. et al. Channels of deformation under conditions of electrostatic stimulation, Metallophysics, 1991, 13[11] 9-13.

[45] Ivanov, Yu.F., Gromov, V.E., Kozlov, T.V., Sosnin, O.V., Evolution of localized deformation channels in a process of electrostimulated drawing of low-carbon steel, Proceedings of Higher Schools - Ferrous Metallurgy, 1997, 6, 42-45.

[46] Speich, G., Swann, P.R., Yield strength and transformation substructure of quenched iron-nickel alloys, J.Iron and Steel Inst., 1965, 203[4] 480-485.

[47] Belous, M.V., Cherepin, V.T., Vasiliev, M.A., Transformations during the Tempering of Steel, Metallurgy, 1973, 232pp.

[48] Belous, M.V., Distribution of carbon by state during tempering of hardened steel, Metallophysics, Rep. Interdepartmental Sat., 1970, 32, 79-82.

[49] Belous, M.V., Shatalova, L.A., Sheiko, Yu.P., The state of carbon in tempered and cold-worked steel. First transformation on vacation, FMM, 1994, 78[2] 99-106.

[50] Belous, M. V., Moskalenok, Yu. N., Cherepin, V.T.,, Sheiko, Yu.P., The state of carbon in tempered and cold-worked steel. Volume effects during heating of hardened Fe-C alloys, FMM, 1995, 80[3] 103-114.

[51] Belous, M.V., Novozhilov, V.B., Shatalova, L.S., Sheiko, Yu.P., Distribution of carbon by state in tempered steel, FMM, 1995, 79[4] 128-137.

[52] Izotov, V.I., Kozlova, A.G., Distribution of carbon in a package of martensitic crystals and its effect on the strength of hardened low-alloy steels, Phys., 1995, 80[1] 97-111.

[53] Izotov, V.I., Filippov, G.A., Influence of supercooling during normal transformation on the distribution of carbon in the ferrite of low-alloy steel, FMM, 1999, 87[4] 72-77.

[54] Speich, G.R., Tempering of low-carbon martensite, Trans. Met. Soc. AIME, 1969, 245[10] 2553-2564.

[55] Kalich, D., Roberts, E.M., On the distribution of carbon in martensite, Met. Trans.,

1971, 2[10] 2783-2790. https://doi.org/10.1007/BF02813252

[56] Fasiska, E.J., Wagenblat, H., Dilatation of alpha-iron by carbon, Trans. Met. Soc. AIME, 1967, 239[11] 1818-1820.

[57] Ridley, N., Stuart, H., Zwell, L., Lattice parameters of Fe-C austenite at room temperature, Trans. Met. Soc. AIME, 1969, 246[8] 1834-1836.

[58] Veselov, S.I., Spektor, E.Z., Dependence of austenite lattice parameter on carbon content at high temperatures, Physical Metallurgy and Metal Science, 1972, 34[5] 895-896.

[59] Lakhtin, Yu.M., Metallurgy and Heat Treatment of Metals, Metallurgiya, 1977, 407pp.

[60] Thomas, G., Sarikaya, M., Lath martensites in carbon steels - are they bainitic?, Proc. Int. Conf. Solid-Solid Phase Transform., Pittsburgh, Pa, Aug. 10-14, 1981, Warrendale, Pa, 1982, 999-1003. https://doi.org/10.2172/7045412

[61] Sarikaya, M., Thomas, G., Steeds, J.W. et al., Solute element partitioning and austenite stabilization in steels, Proc. Int. Conf. Solid-Solid Phase Transform., Pittsburgh, Pa, Aug. 10-14, 1981, Warrendale, Pa, 1982, 1421-1425. https://doi.org/10.2172/7031961

[62] Ivanov, Yu.F., Gladyshev, S.A., Popova, N.A., Kozlov, E.V., Interaction of carbon with defects and processes of carbide formation in structural steels, Transact. "Interaction of Defects of Crystal Lattice and Properties", TulPI, Tula, 1986, 100-105.

[63] Barnard, S.J., Smith, G.D.W., Sarikaya, M., Thomas, G., Carbon atom distribution in a dual phase steel: an atom probe study, Scripta Met., 1981, 15[4] 387-392. https://doi.org/10.1016/0036-9748(81)90216-7

Physics of Strain Hardening of Structural Steels   Materials Research Forum LLC
Materials Research Foundations **153** (2023)   https://doi.org/10.21741/9781644902776

Chapter 4. Hardening mechanisms of bainitic steels

The hardening of metallic materials usually involves three fundamental types: 1. solid-solution hardening (substitutional or interstitial atoms, structural vacancies, short and long-range order, antiphase domains, etc.), 2. sub-structural hardening governed by linear and planar defects, 3. multiphase hardening (carbides and inclusions of retained austenite in steels, break-down of eutectic, composites, etc.). The hardening caused by radiation and quenching defects (vacancies, thermal- and radiation-induced, intrinsic interstitial atoms, etc.) can be included with these three types [1–14].

Within the last couple of decades special attention has been paid to the quantitative evaluation of various physical properties of steels, and considerable advances have been made in understanding their mechanical properties on the basis of the analysis of microstructures [1, 7, 9, 12, 14]. Special attention has been paid to the problem of strength and special features of strength can at present be predicted in many cases, with sufficient reliability, on the basis of alloy composition and microstructure [1, 14]. These results are often obtained using physical models which describe the hardening phenomenon but, in some cases, they can be based upon empirical or semi-empirical assumptions; especially when it is necessary to describe the properties on the basis of an analysis of complicated microstructures of martensite or bainite.

In order to use more efficiently the characteristic strength of steels and, at the same time, obtain the optimum combination of the properties required for efficient application, special importance is attached to understanding the mechanism of hardening in steels. In addition to these considerations, it is important to understand the factors controlling this mechanism and their effect upon many other properties, especially toughness and ductility.

Below are presented the results of numerical evaluations of the hardening mechanisms of the steel, based upon quantitative analysis of the structural parameters of the steel in relation to the strain; i.e. the evolution of the hardening mechanisms of a constructional bainitic steel at different stages of its deformation.

The results of an examination of the structure of bainitic steel, presented in chapter 3, were used for the evaluation of the contributions made by the following mechanisms to the inhibition of moving dislocations: deceleration at forest dislocations, cementite particles or intraphase boundaries and deceleration by interaction with internal stress fields. Estimations of the contributions made by individual hardening mechanisms will be made by using traditional expressions which have been verified for the structure of the steel. The total strength of the steel can be evaluated by using additive and quadratic (with respect to equal-strength obstacles) combinations of the contributions.

Estimates of the contributions of the individual hardening mechanisms and of the total strength of the steel were obtained for the states formed in different stages of the strain-hardening of the steel (see chapter 3). It was therefore possible to analyze the evolution of the hardening mechanisms of the steel and of the overall strength of the steel as a function of the strain.

Two types of boundary are currently recognized: non-intersected (high-angle grain boundaries, boundaries of packets, martensite or bainite crystals) and boundaries intersected by gliding dislocations (low-angle boundaries of sub-grains, martensite or bainite crystals) [15–23].

As mentioned in chapter 3, the α→γ bainitic (intermediate) transformation results in the formation of a plate-shaped structure. The ferrite crystals have very variable dimensions: the longitudinal dimensions change from grain-size and measure from tens of micrometres to units of micrometres, while the transverse dimensions range from units to tenths of micrometres. The ferrite plates are fragmented; i.e., are divided into regions separated by low-angle boundaries. Deformation of steel up to fracture of the specimens does not lead to any change in the dimensions of the grains and ferrite plates but has a strong effect upon the dimensions of the fragments. An increase in strain results in a decrease in the main longitudinal dimensions of the fragments (the transverse dimensions of the fragments are restricted by the boundaries of the ferrite plates and remain almost constant during deformation) (Fig. 4.1, curve 1).

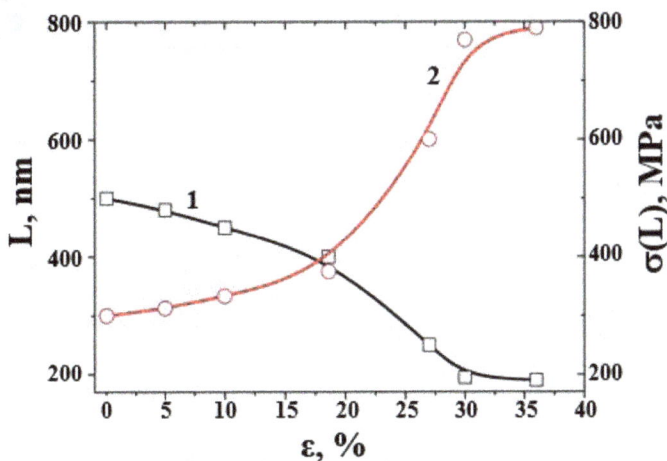

Fig. 4.1. *Effect of strain ε upon the average longitudinal dimension of fragments L (curve 1) and the contribution of the boundaries of fragments σ(L) to the flow stress (curve 2).*

The strain-dependence of the contribution, arising from fragment boundaries, to the strain-hardening of steel with a bainitic structure is shown in Fig. 4.1, curve 2. The following values of the parameters of equation (1) were used in the calculations: L, the average longitudinal dimension of the fragments, k = 0.015, m = 1. It can be seen that increasing deformation of the steel increases the hardening by fragment boundaries from 330MPa to 790MPa (Fig. 4.1, curve 2). This is governed by the decrease in the average size of the fragments (Fig. 4.1, curve 1).

The stress required to sustain plastic deformation, i.e. the flow stress σ, is related to the dislocation density by [1, 6–8, 24–28]:

$$\sigma = \sigma_0 + k\sqrt{\rho}$$

where σ_0 is the flow stress of non-dislocation origin (i.e., stress caused by other hardening mechanisms), ρ is the average (scalar) density of dislocations, $k = maGb$, m is the Schmidt orientation parameter, a is a parameter which characterizes the intensity of inter-dislocation interactions and ranges from 0.1 to 0.51 [22, 29], G is the shear modulus and b is the Burgers vector of the dislocations [30–32].

In ferrite plates, formed as a result of the shear (martensitic) transformation mechanism, there is a network-type dislocation sub-structure with a relatively high scalar density of dislocations, $\approx 7 \times 10^{10}\text{cm}^{-2}$. The plastic deformation of steel is accompanied by an increase in the scalar dislocation density (Fig. 4.2, curve 1). The type of dislocation sub-structure does not change.

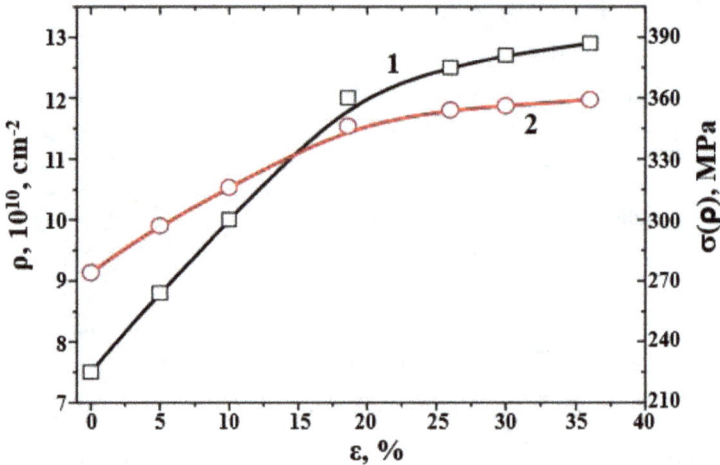

Fig. 4.2. *Dependence of the scalar density of dislocations ρ (curve 1) and the contribution to the flow stress of forest dislocations σ (ρ) (curve 2) upon strain ε.*

Curve 2 in Fig. 4.2 shows the dependence of the contribution, determined by the scalar density of the dislocations, to the strain-hardening of quenched steel with a bainitic structure as a function of the strain. It can be clearly seen that, with increasing strain, the value of the contribution increases in proportion to the increase in the scalar dislocation density and ranges from 280MPa to 360MPa.

The field of internal stresses, generated by a set of material defects [7, 8, 30, 31, 33–35] is an important characteristic as seen from the viewpoint of the movement and multiplication of dislocations and, consequently, of analyses of the strain-hardening mechanisms of the material.

As reported in chapter 2, the procedure for evaluating the magnitude of the internal stress fields is reduced to a determination of the gradient of the curvature–twisting of the crystal lattice [18, 36–44]:

$$\chi = \frac{\partial \varphi}{\partial l} = \frac{0.017}{h}$$

where h is the transverse dimension of the bend extinction contour.

One furthermore estimates the excess density of dislocations as $\rho_\pm = \rho_+ + \rho_-$ (ρ_+ and ρ_- are the densities of positively and negatively charged dislocations, respectively) [18, 19]:

$$\rho_\pm = \frac{1}{b} \cdot \frac{\partial \varphi}{\partial l}$$

The magnitude of the long-range field of the internal stresses is estimated on the basis of [20]:

$$\sigma(h) = \alpha_C Gb \sqrt{\rho_\pm} = \alpha_C Gb \sqrt{\frac{1}{b} \cdot \frac{\partial \varphi}{\partial l}} = \alpha_C G \sqrt{\frac{0.017 \cdot b}{h}},$$

where $\alpha_C = 1$ is the Strunin coefficient [21] and h is the average transverse dimension of the bend extension contour.

Fig. 4.3 shows the strain-dependence of steel with a bainitic structure and the magnitude of the contribution to the flow stress arising from internal (long-range) stress fields.

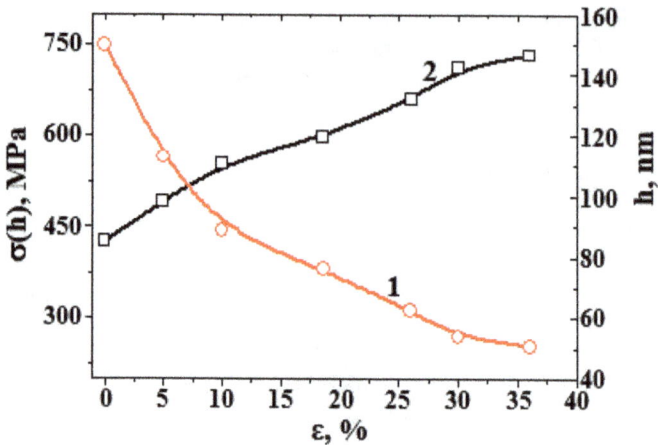

Fig. 4.3. Dependence of the average transverse dimension of the contours h (curve 1) and of the contribution to the flow stress of long-range internal stress fields (curve 2) upon strain ε.

The investigations carried out in this work show that the average transverse dimensions of the contour decrease with increasing straining of the steel (Fig. 4.3, curve 1). According to the equation, the magnitude of the long-range fields of the internal stresses increases (Fig. 4.3, curve 2) as a result of an increase in the curvature-twisting of the crystal lattice of the steel due to the incompatibility-of-deformation of the crystals of bainite, of the grains and of carbide-phase particles.

With regard to the subject of this book, of special interest is the examination of cases of dispersion-hardening by second-phase precipitates and inclusions which can be termed, from the physical viewpoint, as being hardening by coherent and non-coherent particles [45–56] (see chapter 2).

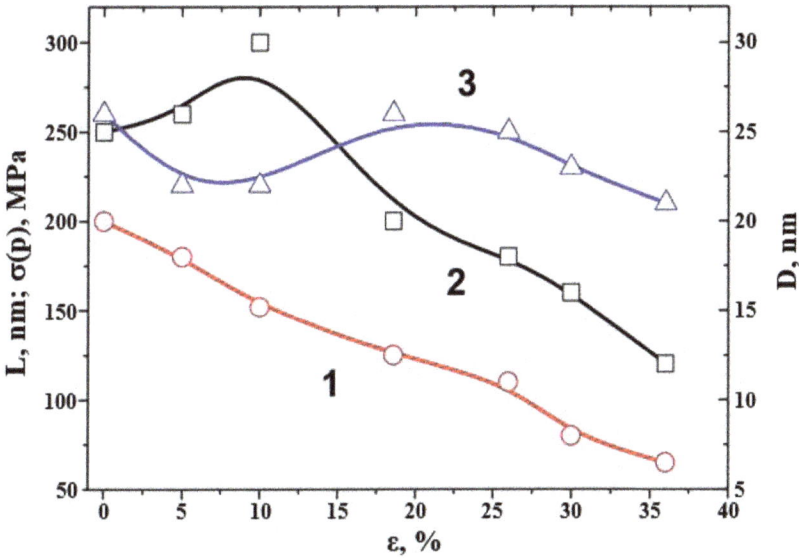

Fig. 4.4. *Dependence of the longitudinal L (curve 1) and transverse D (curve 2) dimensions of cementite particles, and of the contribution to the flow stress of cementite particles σ(p) (curve 3), upon strain.*

The results of analyses carried out in this work show that a characteristic of bainitic steel is the presence of cementite particles. The cementite particle-size prior to deformation of the steel is greater than D_{cr} (Fig. 4.4, curves 1 and 2). Evaluation of the hardening of the steel during deformation, taking into account the presence of the cementite particles, should consequently be carried out by using the relationships obtained for non-coherent precipitates. Curve 3 in Fig. 4.4 shows the dependence of the contribution to the flow stress, arising from the cementite particles, upon the degree of strain of the bainitic steel. It can be clearly seen that this contribution changes in a complicated manner, varying from 210MPa to 260MPa under repeated precipitation of the cementite particles during deformation of the steel.

It is well-known that the addition of substitutional atoms to iron increases the strength of the iron [1, 7, 19, 57–71]. The magnitude of the hardening due to this solid solution depends upon a number of factors (see chapter 2).

The concentration of carbon atoms in the crystal lattice of ferrite was determined, using X-ray diffraction analysis, on the basis of the change in the lattice parameter of the α-phase [2, 28]. The results presented in Fig. 4.5, curve 1, show the increase in the lattice

Materials Research Forum LLC

https://doi.org/10.21741/9781644902776

parameter of the α-phase with increasing strain. This reflects the incorporation of carbon atoms into the crystal lattice of the α-phase.

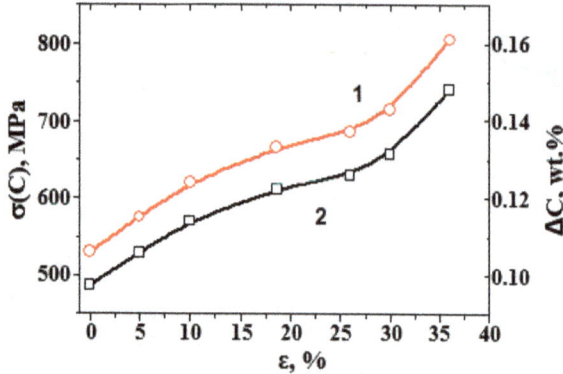

Fig. 4.11. *Strain dependence of the concentration of the carbon atoms within a crystal lattice based upon α-Fe (curve 1) and of the contribution to the flow stress of solid-solution hardening σ(C) (curve 2).*

This result is in good agreement with the data obtained from examination of the evolution of cementite particles during deformation of the steel (Fig. 4.4). The decrease in the particle-size of cementite at high strains may indicate their dissolution, and migration of the carbon atoms to the lattice defects of the steel (dislocations, sub-boundaries and boundaries) and to the solid solution based upon the α-phase.

Enrichment of the crystal lattice of the α-phase with carbon results in a hardening of the steel, with the degree of hardening being governed by the equation. The results presented in Fig. 4.5, curve 2, show that, with increasing straining of the steel, the magnitude of this contribution decreases over a range of 490MPa to 740MPa, as determined by dissolution of the cementite particles, incorporation of some of the carbon atoms into the crystal lattice of iron and their precipitation on dislocations.

Fig. 4.6 shows the results of the evolution of the magnitudes of the contributions to the strain-hardening of constructional bainitic steel, as used for a comparative analysis of their relative values. It can be clearly seen that a relatively large contribution to the hardening of the steel is provided by solid-solution hardening (Fig. 4.6, curve 2), the internal stress fields (Fig. 4.6, curve 3) and, during the final stage of deformation, sub-structural hardening (hardening due to intraphase boundaries) (Fig. 4.6, curve 1). A comparatively small hardening is offered by the dislocation sub-structure (Fig. 4.6, curve 4) of carbide-phase particles (Fig. 4.6, curve 5).

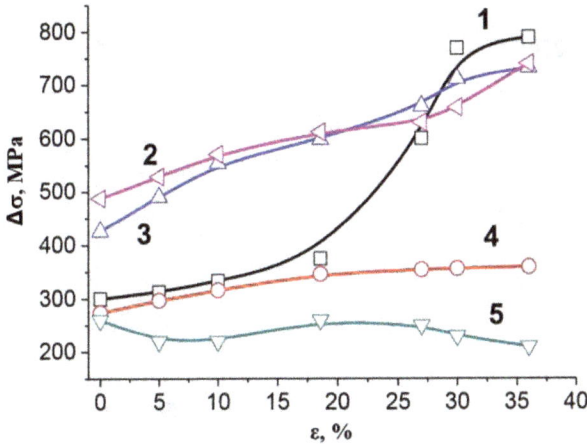

Fig. 4.6. *Strain dependence in a bainitic steel of the contribution to the flow stress arising from intraphase boundaries (1), solid-solution hardening (2), internal stress fields (3), the dislocation sub-structure (4) and cementite particles (5).*

As shown previously the magnitude of the strain-hardening of the steel is determined by the presence of a number of factors. It is assumed that the general yield strength of the steel can be described by a linear sum of the contributions made by the individual hardening mechanisms [1, 19, 69, 72–77]:

$$\sigma = \Delta\sigma_0 + \Delta\sigma_{ip.} + \Delta\sigma_{dis} + \Delta\sigma_{op} + \Delta\sigma_{a.e.} + \Delta\sigma_f$$

where $\Delta\sigma_0$ is the contribution due to friction of the matrix lattice, $\Delta\sigma_{ip}$ is due to intraphase boundaries, $\Delta\sigma_{dis}$ is due to the dislocation sub-structure, $\Delta\sigma_{op}$ is due to the presence of carbide-phase particles, $\Delta\sigma_{a.e.}$ is due to the atoms of alloying elements and $\Delta\sigma_f$ is due to long-range stress fields.

Fig. 4.7 shows strain-hardening curves of steel with a bainitic structure as calculated on the basis of an evaluation of the hardening mechanisms (Fig. 4.7, curve 1) and as determined by experiment (Fig. 4.7, curve 2). It can be seen that the σ–ε dependence obtained from analysis of the hardening mechanisms of the steel (Fig. 4.7, curve 1), at strains exceeding 15%, is considerably greater than that determined by experiment (Fig. 4.7, curve 2). With increasing strain the difference in the experimental and theoretically calculated strain-hardening curves of the steel becomes greater.

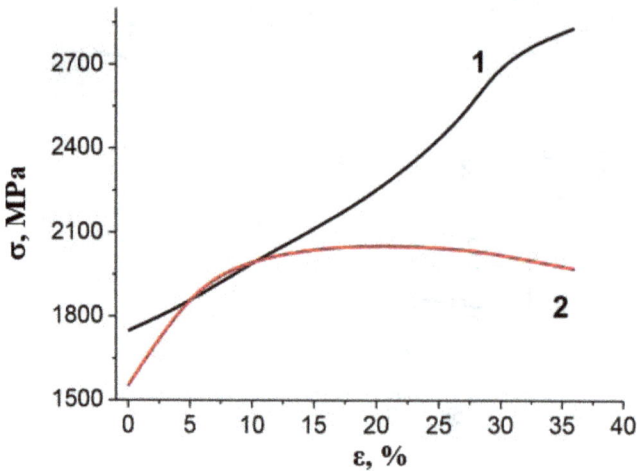

Fig. 4.7. *Strain hardening curves of a steel calculated by evaluating the hardening mechanisms (curve 1) and determined by experiment (curve 2).*

In experiments on Hadfield steel [76, 77] it was established that the inflection-point on the flow curve, leading to a decrease in the strain-hardening coefficient, coincides with the activation of twinning during deformation of the material. Detailed electron-microscopic and micro-diffraction studies permitted the authors of these investigations to conclude that mechanical micro-twinning results in an orientation-softening of the steel and facilitates dislocation glide.

The electron-microscopic studies of bainitic steels carried out in the present work also revealed the occurrence of deformation micro-twinning. Characteristic images of the structure of deformed steel with micro-planes are shown in Fig. 4.8a. At strains of $\varepsilon = 5$ and 10%, strain-twinning of the steel is not pronounced (Fig. 4.8b). At higher values of ε the volume fraction of the material which is affected by strain micro-twinning greatly increases. Taking into account the results obtained in [76, 77] it can consequently be assumed that the differences in the experimentally determined and theoretically calculated strain-hardening curves, which were most pronounced at high strains, are due to the onset of micro-twinning during deformation of the steel.

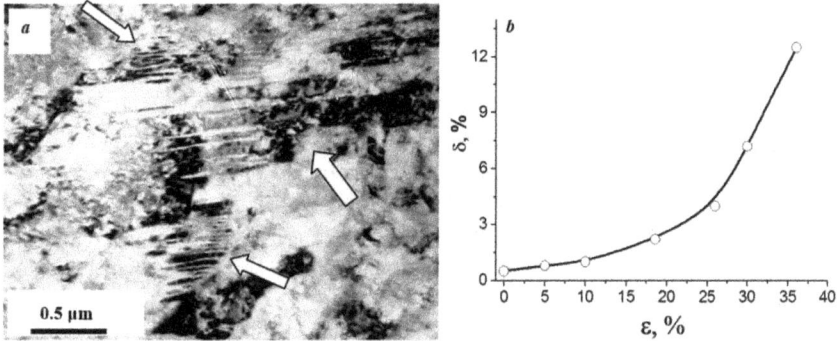

Fig. 4.8. *Electron microscopic image of the structure of steel following deformation ε = 30% (a); b) strain dependence of the volume of material containing micro-twins.*

Chapter 4 Conclusions

The results of a quantitative analysis of the structure of bainitic steel subjected to uniaxial compression were used to evaluate the hardening mechanisms. Analysis of the nature of the strain-hardening shows that:

1. Hardening of the investigated steel depends upon a large number of factors.

2. The largest contribution to strain-hardening of the investigated steel is provided by sub-structural hardening (hardening due to long-range internal stress fields and to fragmentation of the structure) and solid-solution hardening, determined by the incorporation of carbon atoms into the crystal lattice of ferrite;

3. It is assumed that the softening of steel with a bainitic structure at high (greater than 15%) strains is due to the onset of deformation micro-twinning.

Chapter 4 References

[1] Pickering, F.B., Physical Metallurgy of Steels and Development, Metallurgiya, 1982, 184pp.

[2] Kelly, A., Nicholson, R.B., Strengthening Methods in Crystals, Elsevier, 1971, 214pp.

[3] Fleischer, R.L., Hibbard, W.R., The Relation between the Structure and Mechanical Properties of Metals, HMSO, 1963, 203pp.

[4] Smirnov, B.I., Dislocation Structure and Hardening of Crystals, Nauka, Leningrad, 1981, 236pp.

[5] Trefilov, V.I., Milman, Yu.V., Firstov, S.A., Physical Basis of the Strength of the Refractory Metals, Naukova Dumka, Kiev, 1975, 315pp.

[6] Trefilov, Yu.I., Moiseev, V.I., Pechkovskii, E.P. et al. Deformation and Fracture Consolidation of Polycrystalline Metals, Naukova Dumka, Kiev, 1987, 248pp.

[7] Shtremel, M.A., The Strength of Alloys: Part II. Deformation. Textbook for High Schools, Moscow Institute of Steel and Alloys, Moscow, 1997, 527pp.

[8] Koneva, N.A., Kozlov, E.V., Physical Nature of Stage Character of Plastic Deformation, Structural Levels of Plastic Deformation Fracture (Ed. V.E.Panin), Nauka, Siberian branch, Novosibirsk, 1990, 123-186.

[9] Gromov, V.E., Kozlov, E.V., Bazaikin, V.I., Tsellermayer, V.Ya., Ivanov, Yu.F., et al., Physics and Mechanics of Drawing and Forging, Nedra, Moscow, 1997, 293pp.

[10] Vladimirov, V.I., Physical Theory of Strength and Ductility. Point Defects. Strengthening and Recovery, LPI, Leningrad, 1975, 120pp.

[11] Rybin, V.V., Large Plastic Deformation and Fracture of Metals, Metallurgiya, 1986, 224pp.

[12] Finkel, V.M., Physical Fundamentals of Inhibition of Failure, Metallurgiya, 1977, 359pp.

[13] Kaybishev, O.V., Valiev, R.Z., Grain Boundaries and Properties of Metals, Metallurgiya, 1987, 216pp.

[14] Statistical Strength and Mechanics of Steel Fracture: Transaction of Scientific Papers (Translated from German), Eds. V.Dal & V.Anton, Metallurgiya, 1986, 566pp.

[15] Hall, E.O., The deformation and ageing of mild steel: III discussion of results, Proc. Phys. Soc., 1951, 64B, 747-753. https://doi.org/10.1088/0370-1301/64/9/303

[16] Petch, N.J. The cleavage strength of polycrystals, J.Iron Steel Inst., 1953, 174, 25-28.

[17] Luke, K., Gottshteing, G., Atomic mechanisms of metal plasticity, Statistical strength and mechanics of steel fracture, Transaction (Translated from German), Eds. V.Dal & V.Anton), Metallurgiya, 1986, 14-36.

[18] Dal, V., Increase in strength at the expense of grain refinement, Statistical strength and mechanics of steel fracture: Transaction. (Translated from German), Eds. V.Dal & V.Anton), Metallurgiya, 1986, 133-146.

[19] Goldstein, M., Farber, B.M., Precipitation Hardening of Steel, Metallurgiya, 1979, 208pp.

[20] Belenkiy, B.Z., Farber, B.M., Goldshtein, M.I., Estimation of strength of low-carbon low-alloy steels by structural data, FMM, 1975, 39, 403-409.

[21] Naulor, I.R., The influence of the lath morphology on the yield strength and transition temperature of martensite-bainite steel, Met. Trans., 1979, 10A[7] 873-

891. https://doi.org/10.1007/BF02658305

[22] McLean, D., Mechanical Properties of Metals, Metallurgiya, 1965, 431pp.

[23] Ashby, M.F., Mechanisms of deformation and fracture, Adv. Appl. Mech., 1983, 23, 118-177. https://doi.org/10.1016/S0065-2156(08)70243-6

[24] Keh, A.S., Direct Observations of Crystals, Interscience, 1962, 213pp.

[25] Bailey, J.E., Hirsch, P.B., The dislocation distribution, flow stress and stored energy in cold-worked polycrystalline silver, Phil. Mag., 1960, 53, 485-497. https://doi.org/10.1080/14786436008238300

[26] Kuhlman-Wilsdorf, D., A critical test of theories of work-hardening for the case of drawn iron wire, Met. Trans., 1970, 1, 3173-3179. https://doi.org/10.1007/BF03038434

[27] Predvoditelev, A.A., The state-of-the-art of studying dislocation ensembles, Problems of Modern Crystallography, Nauka, 1975, 262-275.

[28] Lavrentev, F.F., The type of distribution as the factor determining work hardening, Mat. Sci. Eng., 1980, 16, 191-208. https://doi.org/10.1016/0025-5416(80)90175-5

[29] Embyri, I.D., Strengthening by dislocations structure, Strengthening Method in Crystals, Applied Science Publishers, 1971, 331-402.

[30] Koneva, N.A., Kozlov, E.V., Nature of substructural strengthening, Proceedings of Higher Schools - Physics, 1982, 8, 3-14. https://doi.org/10.1007/BF00895238

[31] Koneva, N.A., Kozlov, E.V., Physics of substructural stress strengthening, Bulletin TGASU, 1999, 1, 21-35.

[32] Kocks, U.F., Statistical treatment of penetrable obstacles, Canadian Journal of Phys., 1967, 45[2] 737-755. https://doi.org/10.1139/p67-056

[33] Strunin, B.M., Probabilistic description of internal stress field with random arrangement of dislocations, FTT, 1971, 13[3] 923-926.

[34] Seeger, A., Mechanism of gliding and strengthening in fcc and hexagonal close-packed metals, Dislocations and Mechanical Properties of Crystals, IIL, 1960, 179-289.

[35] Hirt, J., Lotte, I., Theory of Dislocations, Atomizdat, 1972, 599pp.

[36] Ivanov, Yu.F., Kozlov, E.V., Electron-microscopic analysis of martensite phase of the 38CrNi3MoV steel, Proceedings of Higher Schools - Ferrous Metallurgy, 1991, 8, 38-41.

[37] Koneva, N.A., Lychagin, D.V., Teplyakova, L.A., Kozlov, E.V., Dislocation-disclination substructures and hardening, Theoretical and Experimental Study of Dislocations, FTI, 1984, 116-126.

[38] Vladimirov, V.I., Physical Theory of Strength and Plasticity. Point Defects.

Materials Research Forum LLC
https://doi.org/10.21741/9781644902776

Strengthening and Recovery, LPI, 1975, 120pp.

[39] Eshelby, J., Continuum Theory of Dislocations, TLI, 1963, 247pp.

[40] Shtremel, M.F., Hardness of Alloys: Part I. Lattice Defects, MISIS, 1999, 384pp.

[41] Hirsch, P.B., Howie, A., Nicholson, R.B. et al. Electron Microscopy of Thin Crystals, Mir, 1968, 577pp.

[42] Koneva, N.A., Kozlov, E.V., Trishkina, L.I., Lychagin, D.V., Long-range stress fields, curvature-torsion of crystal lattice and stages of plastic deformation. Methods of measurement and results, New Methods in Physics and Mechanics of Deformed Solid, Transactions of international conference, TGU, Tomsk, 1990, 83-93.

[43] Koneva, N.A., Lychagin, D.V., Teplyakova, L.A., Kozlov, E.V., Turns of crystal lattice and stages of plastic deformation, Experimental Investigation and Theoretical Description of Dislocations, FII, 1984, 161-164.

[44] Teplyakova, L.A., Ignatenko, L.N., Kasatkina, N.F., Ivanov, Yu.F. et al., Regularities of plastic deformation of steel with a structure of tempered martensite, Plastic Deformation of Alloys. Structurally-Inhomogeneous Materials, TGU, Tomsk, 1987, 26-51.

[45] Hornbogen, E., Increase of hardness by disperse precipitates: Transaction. (Translated from German), Eds. V.Dal & V.Anton, Metallurgiya, 1986, 165-189.

[46] Orowan, E., Symposium on Internal Stresses in Metals and Alloys, Inst. Metals, London, 1948, 451pp.

[47] Tekin, E., Kelly, P.M., Tempering of Steel Precipitation from Iron Based Alloys, Gordon & Breach, 1965, 283pp.

[48] Ashby, M.F., Physics of Strength and Plasticity, MIT Press, Cambridge, Mass., 1969, 113pp.

[49] Eshelby, J. D., The stresses at the inclusion-matrix interface, Progress in Solid Mechanics, Chap. 3, Vol. 2, Interscience, Wiley, New York, 1961, 534-541.

[50] Ansell, G.S., Lenel, F.V., Criteria for yielding of dispersion-strengthened alloys, Acta Met., 1960, 8[9] 612-616. https://doi.org/10.1016/0001-6160(60)90015-8

[51] Hirsch, P.B., Humphreys, F.J., Plastic deformation of two-phase alloys containing small non-deformable particles, Physics of Strength and Plasticity, Metallurgiya, 1972, 158-186.

[52] Kelly, A., Nicholson, R., Dispersion Hardening, Metallurgiya, 1966, 187pp.

[53] Pyabko, P.V., Ryaboshapka, K.P., Theories of yield point of heterophase systems with coherent particles, Metallophysics, 1970, 31, 5-31.

[54] Gerold, V., Habercorn, H., On the critical resolved shear stress of solid solutions

containing coherent precipitates, Phys. Status Solidi, 1966, 16[2] 675-684. https://doi.org/10.1002/pssb.19660160234

[55] Fleischer, R.L., Dislocation structure in solution hardened alloys, Electron Microscopy and Strength of Crystals, Interscience, New York, 1963, 973-989.

[56] Mott, N.F., Nabarro, F.R.N., The distribution of dislocations in slip bands, Proc. Phys. Soc., 1940, 52[1] 86-93. https://doi.org/10.1088/0959-5309/52/1/312

[57] Friedel, G., Dislocations, Mir, 1967, 643pp.

[58] Krishtal, M.A., Interaction of dislocations with impurity atoms and properties of metals, Physics and Chemistry of Material Treatment, 1975, 1, 62-71.

[59] Fleischer, R.L., Substitutional solution hardening, Acta Met., 1963, 11, 203-209. https://doi.org/10.1016/0001-6160(63)90213-X

[60] Pickering, O.F.B., Gladman, T., Iron and Steel Inst. Spec. Rep., 1963, 81, 10.

[61] Dyson, D.J., Holmes, B., J. Iron Steel Inst., 1970, 208, 469.

[62] Fleischer, R.L., Hibbard, W.R., The Relation Between the Structure and Mechanical Properties of Metals, HMSO, 1963, 261pp.

[63] Fasiska, E.J., Wagenblat, H., Dilatation of alpha-iron by carbon, Trans. Met. Soc. AIME, 1967, 239[11] 1818-1820.

[64] Kalich, D., Roberts, E.M., On the distribution of carbon in martensite, Met. Trans., 1971, 2[10] 2783-2790. https://doi.org/10.1007/BF02813252

[65] Barnard, S.J., Smith, G.D.W., Sarikaya, M., Thomas, G., Carbon atom distribution in a dual phase steel: an atom probe study, Scripta Met., 1981, 15[4] 387-392. https://doi.org/10.1016/0036-9748(81)90216-7

[66] Ridley, T., Stuart, H., Zwell, L., Lattice parameters of Fe-C austenite at room temperature, Trans. Met. Soc. AIME, 1969, 246[8] 1834-1836.

[67] Veselov, S.I., Spektor, E.Z., Dependence of austenite lattice parameter on carbon content at high temperatures, FMM, 1972, 34[5] 895-896.

[68] Vohringer, O., Macherauch, E., Struktur und Mechanische eigenschaft von martensite, HTM, 1977, 32[4] 153-202. https://doi.org/10.1515/htm-1977-320401

[69] Norstrom, L.A., On the yield strength of quenched low-alloy lath martensite, Scandinavian J.Met., 1976, 5[4] 159-165.

[70] Cottrell, A.H., Dislocations and Plastic Flow in Crystals, Metallurgiya, 1958, 267pp.

[71] Cottrell, A.H., Bilby, B.A., Dislocation theory of yielding and strain ageing of iron, Proc. Phys. Soc. A., 1949, 62, 49-53. https://doi.org/10.1088/0370-1298/62/1/308

[72] Prnka, T., Quantitative relations between parameters of disperse precipitates and mechanical properties of steels, Physical Metallurgy and Thermal Treatment of Steel, 1979, 7, 3-8.

[73] Toronen, T., Kotilainen, H., Nehonen, P., Combination of elementary hardening mechanisms in Fe-Cr-Mo-V steel, Proc. Int. Conf. Martensite Trans. ICOMAT-1979, Cambridge, 1979, 2, 1437-1442. https://doi.org/10.1016/B978-1-4832-8412-5.50237-X

[74] Butler, E.R., Burke, M.G., Martensite formation at grain boundaries in sensitised 304 stainless steel, J.Physique, 1982, 43[12] 121-126. https://doi.org/10.1051/jphyscol:1982411

[75] Orowan, E., Conditions for dislocation passage of precipitates, Proc. Symp. Intern. Stress in Metals and Alloys, Inst. Met., London, 1948, 451-454.

[76] Koneva, N.A., Kiseleva, S.F., Popova, N.A., Kozlov, E.V., Evolution of internal stresses and stored energy density during deformation of austenitic steel 110G13, Deformation and Destruction of Materials, 2013, 9, 38-42.

[77] Kiseleva, S.F., Popova, N.A., Koneva, N.A., Kozlov, E.V., Effect of transformation micro-twins on the excess dislocation density and internal stresses of a deformed fcc material. Izv. RAN. Physical series, 2012, 76, 70-74.

Chapter 5. Strain hardening of structural steel with pearlite structure

5.1. Structural and phase changes in pearlitic steel under uniaxial tension

The purpose of this section is to analyze the pearlite defect sub-structure of the lamellar morphology of rail-steel, fractured in uniaxial tension.

Samples of rail-steel were used as the research material. Their properties and elemental composition are regulated by GOST R51685–2000. Mechanical tests were carried out by the uniaxial tension of flat proportional samples having the form of double-sided blades, with the dimensions of the working area of the blades being 1.5 x 4.45 x 8.0mm^3. Uniaxial tensile deformation was carried out using an Instron 3369 testing machine at a loading rate of 1.2mm/min. As in previous chapters the structure of the fracture surface was studied by means of scanning electron microscopy (Philips SEM 515). The defect sub-structure of the steel in the fracture zone was investigated using transmission (thin-foil) electron diffraction microscopy (JEM-2100 JEOL).

Fig. 5.1 shows the machine stress-strain curves obtained by uniaxial tensile testing of three steel samples. The tests showed that the tensile strength varies from 1247MPa to 1335MPa; the deformation of the samples during fracture ranging from 0.69 to 0.75.

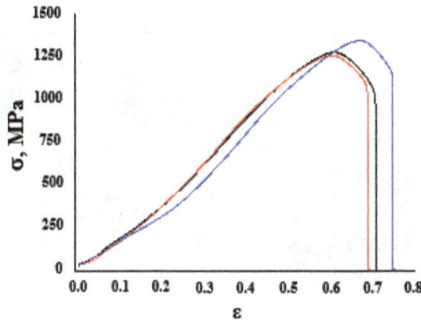

Fig. 5.1. *Stress-strain curves of pearlitic steel subjected to uniaxial tension.*

Characteristic images of the steel fracture surface are shown in Fig. 5.2. As a rule, when samples are deformed in tension three zones are formed at the fracture surface: a fibrous zone (central part of the sample), a radial zone following it and then, along the edge of the sample, a cut zone [1]. The fibrous zone of a smooth rectangular sample is elliptical in shape, with its long axis parallel to the long sides of the rectangle (Fig. 5.2*a*). The radial zone of rectangular samples whose width is much greater than their thickness has the form of a chevron or 'herringbone' pattern (Fig. 5.2a). Chevron patterns are often associated with unstable and relatively rapid crack propagation. The appearance of the chevron pattern is due to a discrepancy between the general direction of crack propagation and the shortest direction from the crack-front to the free surface. In this case radial scars spread towards the free surface, forming chevron patterns [1]. The tops of the V-shaped chevrons are directed away from the fracture source. The focus of the fracture in our case is consequently located on the left-hand edge of the sample (Fig. 5.2a, b).

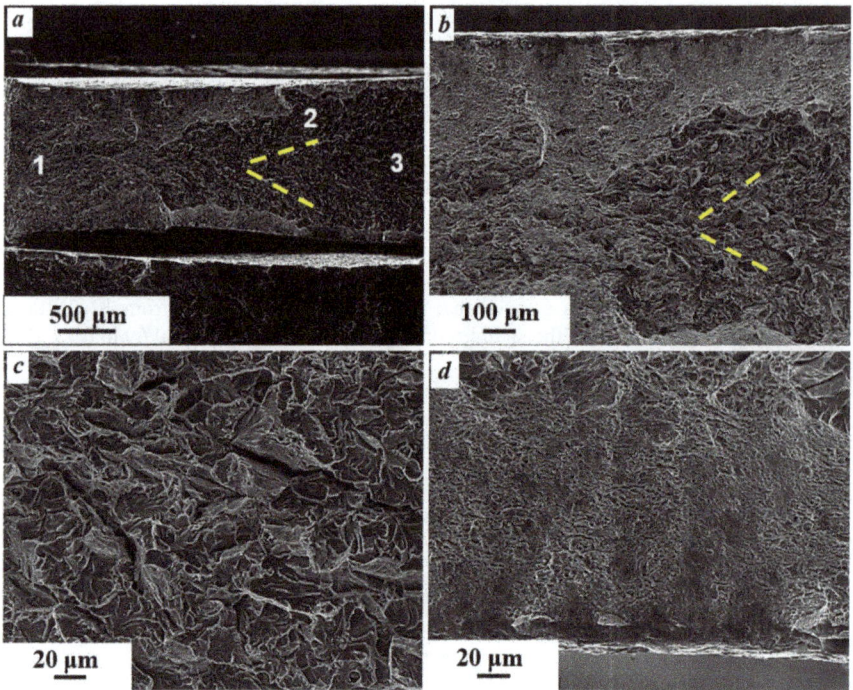

Fig. 5.2. *Fractography of the fracture surface of steel subjected to uniaxial tension: 1 – cut zone, 2 – radial zone, 3 – fibrous zone.*

In [2–4] we showed that the following components are distinguishable in the structure of the studied steel according to their morphological characteristics: grains of pearlite of lamellar morphology, grains of a ferrite-carbide mixture (grains of irregular pearlite) and grains of structure-free ferrite (ferrite grains within the bulk of which there are no particles of carbide phase). The main type of structure of the studied steel comprises grains of lamellar pearlite, the relative content of which is 0.7, while the relative content of grains of ferrite-carbide mixture is 0.27 and the remainder consists of grains of structure-free ferrite.

As a rule, the structure of lamellar pearlite consists of alternating ferrite plates (solid solution based upon the bcc iron lattice) and cementite plates (Fe_3C, iron carbide with an orthorhombic crystal lattice) [5, 6]. The fracture of steel, under uniaxial tension, of flat samples does not lead to a change in the morphology of the material. Grains having the lamellar structure which is characteristic of pearlite are present both in the fracture zone and away from it (Fig. 5.3). Transformation of the steel structure is revealed at the level of the defect sub-system and is accompanied by multiple transformation of pearlite.

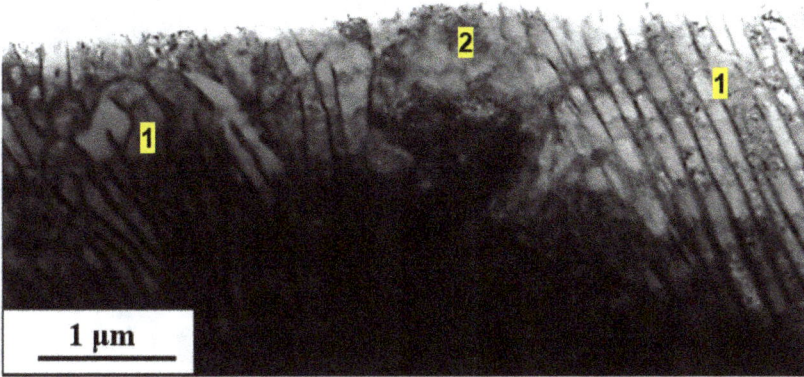

Fig. 5.3. *Electron microscopic image of a steel structure tested to fracture: 1 – grains of lamellar pearlite; 2 – grains of ferrite-carbide mixture.*

It is first necessary to consider the transformation of the ferrite plate structure. It is known that the ferrite plates of pearlite colonies are fragmented; i.e., they are divided into regions which are separated by low-angle boundaries. The fragmentation of ferrite is most clearly seen in dark-field analyses of the structure (Fig. 5.4).

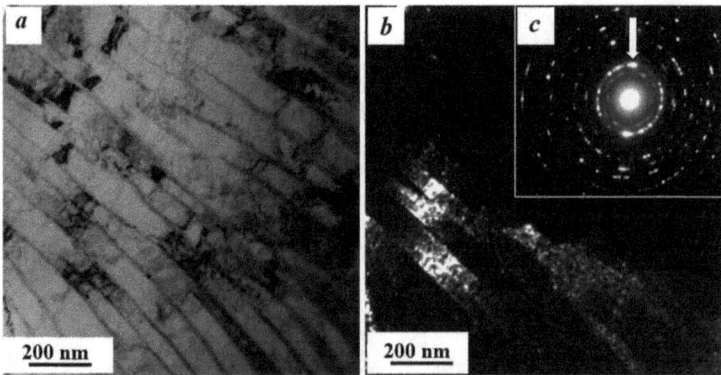

Fig. 5.4. *Electron microscopic image of the structure of P65 steel tested to fracture in uniaxial tension; a – light field b – dark field obtained for the [110] α–Fe reflection; c – electron-microscopic-diffraction pattern, the arrow indicates the reflection for which the dark-field image is obtained (b).*

Deformation is accompanied by the formation of a dislocation sub-structure within the volume of ferrite plates (Fig. 5.5). Dislocations are distributed chaotically, or form clusters. The scalar dislocation density is $7.9 \cdot x \ 10^{10} cm^{-2}$.

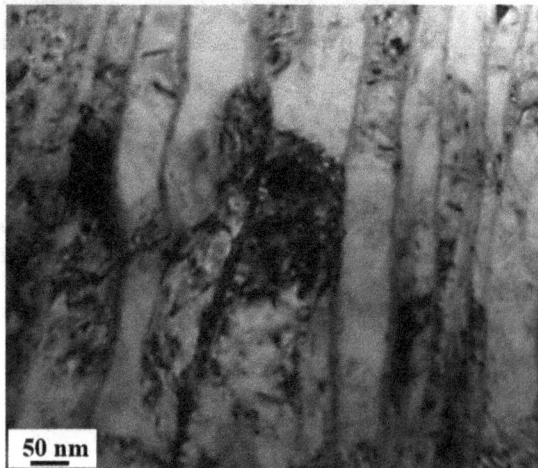

Fig. 5.5. *Electron microscopic image of the dislocation sub-structure of ferrite plates of P65 steel, tested to fracture in uniaxial tension.*

Steel deformation is accompanied by the formation of stress fields within the sample. When studying material via the electron-microscopy of thin foils, the internal stress fields manifest themselves in the form of bend extinction contours, located mainly in ferrite plates. The sources of the stress fields in the steel under study are the boundaries of cementite plates and ferrite plates (Fig. 5.6), as well as grain boundaries (Fig. 5.3). It should be noted that tension of the steel under study is accompanied by a rotation of the pearlite grains, which is most pronounced in the fracture zone of the samples (Fig. 5.6). This suggests the occurrence of a rotational mode of deformation within the fracture zone of the sample.

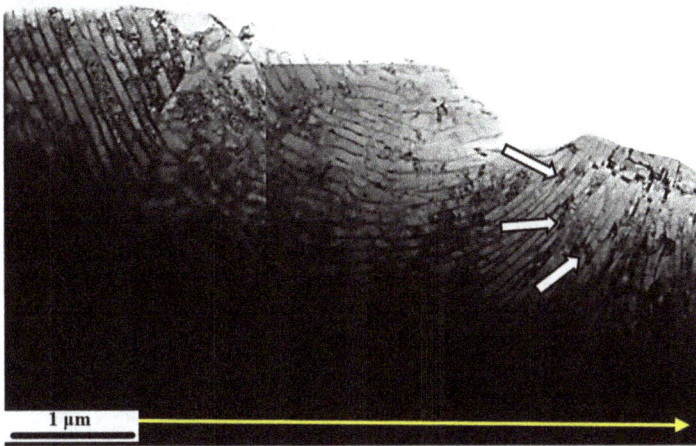

Fig. 5.6. *Electron microscopic image of the structure of steel P65 tested to fracture in uniaxial tension; the arrows indicate the extinction bend contours. The long coloured arrow indicates the tensile stress direction (longitudinal axis of the sample).*

Deformation of the examined steel is accompanied by the fracture of cementite plates. One of the fracture mechanisms is the dissolution of cementite plates [7]. Carbon atoms are carried, by moving dislocations, into the volume of ferrite plates followed by the formation of nanoscale iron carbide particles (Fig. 5.7). The average size of the particles located in the ferrite plates is 8.3nm. Particles of this size are most clearly revealed when using the dark-field technique (Fig. 5.7*b*).

The dissolution of cementite is accompanied by the formation of a region of material around the plates which differs in contrast from the main grain volume (Fig. 5.8*a*). It can be assumed that the change in contrast is due to a change in the chemical composition of the material surrounding the cementite plate, namely an increased carbon concentration.

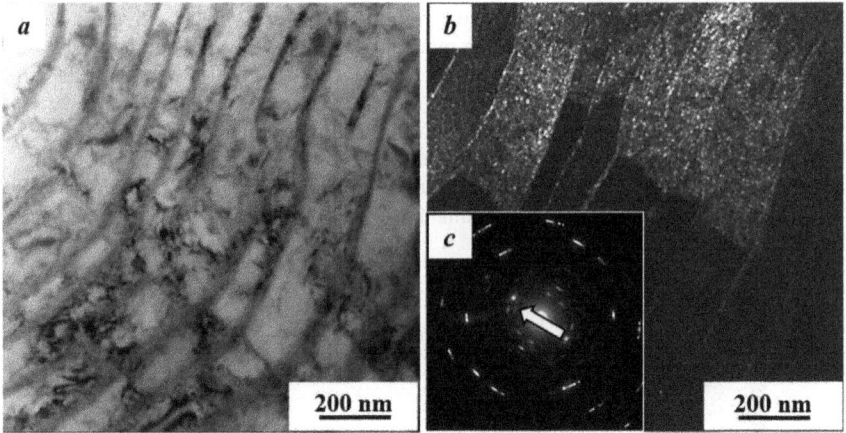

Fig. 5.7. *Electron microscopic image of nanoscale cementite particles formed in the ferrite plates of P65 steel tested to fracture in uniaxial tension; a – light field b – dark field obtained for the [110]α-Fe + [121]Fe₃C reflection; c – electron-microscopic diffraction pattern, the arrow indicates the reflection for which the dark-field image is obtained (b).*

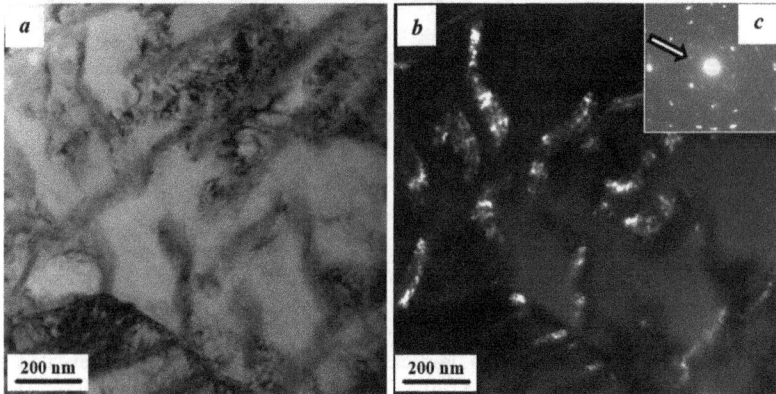

Fig. 5.8. *Electron microscope image of nanoscale cementite particles formed in the ferrite plates of P65 steel tested to destruction in uniaxial tension; a – light field; b – dark field obtained for the [230] Fe₃C reflection; c – electron-microscopic-diffraction pattern, the arrow indicates the reflection for which the dark-field image is obtained (b).*

Together with dissolution, the plastic deformation of steel is accompanied by the fragmentation of cementite plates. It was found that, in the fracture zone of samples, the cementite plates although maintaining their original morphology are divided into regions (regions of coherent scattering) with an average size of 9.3nm (Fig. 5.8, Fig. 5.9).

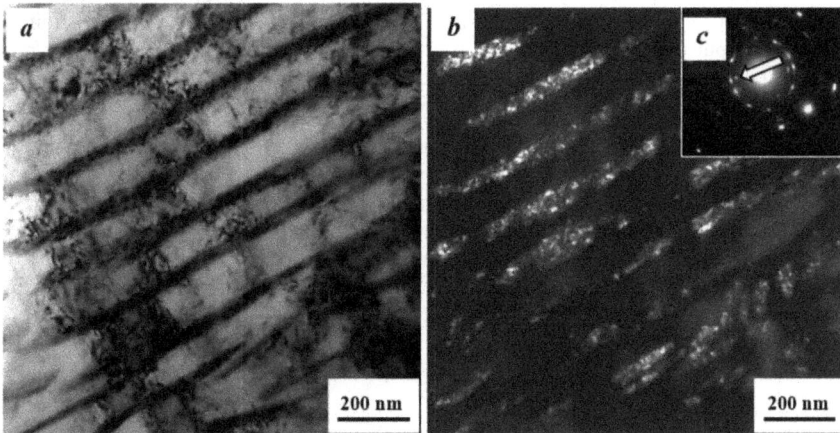

Fig. 5.9. *Electron microscope image of nanoscale cementite particles formed in the ferrite plates of P65 steel tested to destruction in uniaxial tension; a – light field b – dark field obtained for the [110] α-Fe + [121] Fe₃C reflection; c – electron-microscopic diffraction pattern, the arrow indicates the reflection for which the dark-field image is obtained (b).*

The pearlite defect sub-structure of the lamellar morphology of rail-steel subjected to fracture by the uniaxial tension of flat proportional samples was studied. It is known that the ultimate tensile strength varies from 1247MPa to 1335MPa and that the deformation of the samples during fracture ranges from 0.69 to 0.75. It is shown that steel deformation is accompanied by the splitting of ferrite plates, by low-angle boundaries, into fragments and by a significant increase in the scalar density of dislocations to $7.9 \cdot x \ 10^{10} cm^{-2}$ (the scalar density of dislocations in the original steel is $3.2 \cdot x \ 10^{10} cm^{-2}$). The destruction of cementite plates via the dissolution mechanism was revealed followed by the removal of carbon, by moving dislocations, into the bulk of the ferrite plates with the formation of nanoscale (8.3nm) particles of rounded cementite within them. It is shown that the dissolution of cementite plates is accompanied by their fragmentation (separation into coherent scattering regions, the average size of which is 9.3nm).

5.2. Evolution of the structure of lamellar pearlite under uniaxial compression

The purpose of this section is to analyze the evolution of the pearlite defect sub-structure of the lamellar morphology under uniaxial compression.

Samples of rail-steel were used as the research material. Its properties and elemental composition are regulated by GOST R51685–2000. The samples had the shape of a parallelepiped with dimensions of 9.6 x 4.7 x 4.7mm^3. Uniaxial compression straining was carried out using an Instron 3369 testing machine at a loading rate of 1.2mm/min. The structural and phase state of steel subjected to deformation by 15%, 30% and 50% was analyzed.

Fig. 5.10 shows the machine stress-strain curves obtained by the uniaxial compression of pearlitic steel. As a rule, the change in the cross-sectional area of samples during this method of loading is not taken into account and therefore the graph in question should be called a conditional material compression diagram. Because the samples of the steel under study could not be fractured by compression testing (Fig. 5.10) the compressive strength limit could not be determined. That is, the steel could deform greatly without fracturing; the samples being merely flattened.

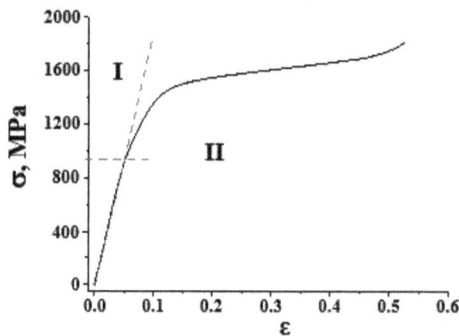

Fig. 5.10. *Stress-strain curves of rail steel under uniaxial compression.*

The strain-hardening diagram of steel is a pronounced parabola (Fig. 5.10). Analysis of the stress-strain curves of metal hardening is based upon the concept of their being stages of strain-hardening which reflect the evolution of the dislocation structure during deformation [8]. In the stress-strain curves (Fig. 5.10) it is possible to distinguish a stage of elastic deformation (stage I) and a stage (II) of plastic deformation having a parabolic functional dependence of the form:

$$\sigma = \sigma_0 + \theta\varepsilon^n \tag{5.1}$$

where σ_0 is the yield stress (equal to 900-930MPa), $\theta(\varepsilon)=d\sigma/d\varepsilon$ (strain-hardening coefficient) and n < 1 is the strain-hardening index [8].

It is shown in [9] that, if one transforms the $\sigma = f(\varepsilon)$ dependence into $d\sigma/d\varepsilon = f(\varepsilon)$ coordinates, one can then reveal the multistage strain-hardening of the material. In most cases the following stages can be distinguished (Fig. 5.11): transitional (N), following the yield strength and reflecting either an increase or decrease in the strain-hardening coefficient. This is immediately followed by stage II, with a high constant, or almost constant, high hardening. In the next stage (III), the strain-hardening coefficient decreases. The $\sigma = f(e)$ dependence in this section is parabolic, or close to it. Finally, stage III is followed by stage IV; with a very low and constant hardening coefficient.

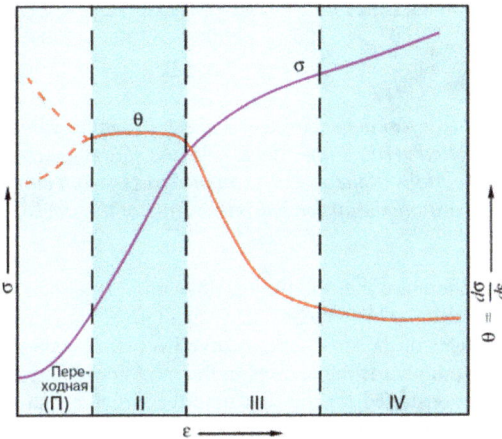

Fig. 5.11. *Typical four-stage stress σ – strain ε curve and $\theta = f(\varepsilon)$ dependence. Dashed lines demarcate the boundaries of the deformation stages [9].*

After processing the deformation curve shown in Fig. 5.10, the strain-hardening stages of steel are revealed in the $d\sigma/d\varepsilon = f(\sigma)$ coordinates (Fig. 5.12). According to the inflection points of the hardening curve, rearranged in these coordinates, the structural changes occurring in the material are analyzed [10].

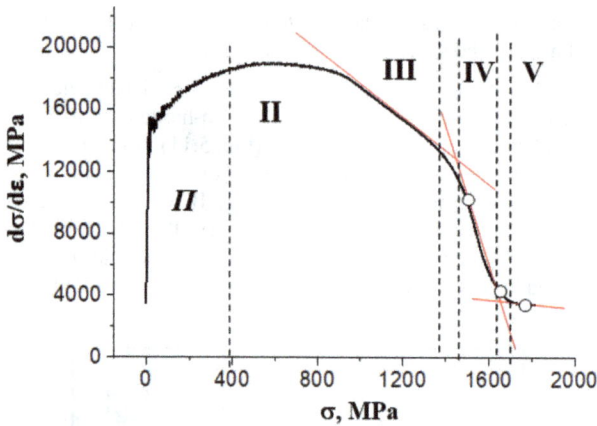

Fig. 5.12. *The dσ/dε = f(σ) dependence of rail steel samples subjected to uniaxial compression. The dotted vertical lines demarcate the stages of strain hardening. The circles indicate the locations of the samples, on the stress-strain curve, which were used to study the structural-phase state of the steel.*

It is known from the literature that the plastic-flow stage is associated with a change in the hardening mechanism, which means that qualitatively different defect structures occur at successive stages of the stress-strain curve [8–10]. The stages are detected when processing the stress-strain hardening curves in the $\sigma{-}\varepsilon^n$ coordinates. In [8] V.I. Trefilov and his colleagues demonstrated for the first time the structural nature of the hardening stages and connected linear sections on the hardening curve processed in $\sigma - e^{0.5}$ coordinates with changes in the structural state of the material. In the P65 steel studied here, the stages of plastic flow of the material are most clearly revealed when processing the strain-hardening curves in $\sigma = f(\varepsilon^{0.2})$ coordinates (Fig. 5.13).

120

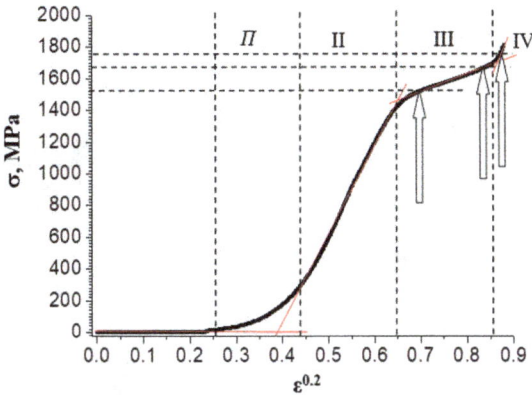

Fig. 5.13. *The σ = f(ε$^{0.2}$) dependence of rail-steel samples subjected to uniaxial compression. The dotted vertical lines demarcate the stages of strain hardening. The arrows indicate the locations, on the stress-strain curve, of the samples used to study the structural-phase state of the steel.*

Analyzing the results presented in Fig.5.12 and Fig.5.13, it is possible to detect the multi-stage nature of the strain-hardening of rail-steel. Samples having a low strain-hardening coefficient were studied (Fig. 5.12, the location of the examined samples on the $d\sigma/d\varepsilon = f(\sigma)$ curve is indicated by circles).

As previously noted the following components can be distinguished in the structure of the studied steel according to their morphological characteristics: pearlite grains of lamellar morphology, grains of ferrite-carbide mixture (grains of irregular pearlite) and grains of structure-free ferrite (grains of ferrite within the volume of which there are no particles of carbide phase). The main structure of the studied steel comprises pearlite grains; the relative content of which is 0.7, while the relative content of ferrite-carbide grains is 0.27 and the remainder 0.03) consists of structure-free ferrite grains. A dislocation sub-structure in the form of chaotically distributed dislocations or, less frequently, dislocation grids is observed in the bulk of all the above structural components of steel. The scalar dislocation density in ferrite grains $<\rho> = 3.2 \cdot x\ 10^{10}$ cm^{-2} while, in pearlite grains, $<\rho> = 4.2 \cdot x\ 10^{10}$ cm^{-2}.

Steel deformation is accompanied by multiple transformations of the steel structure. Material fragmentation is first observed and expands with increasing degree of deformation. At $\varepsilon = 50\%$, the fragmented steel structure occupies 0.37 of the volume of the foil under study. A characteristic electron-microscopic image of the pearlite structure formed at a given degree of deformation is shown in Fig.5.14. Fragments formed in ferrite plates are separated by low-angle boundaries (Fig.5.14*b*, these boundaries are indicated by arrows). With increasing degrees of deformation, the average sizes of ferrite plate fragments decrease from 240nm (ε = 15%) to 200nm (ε = 50%).

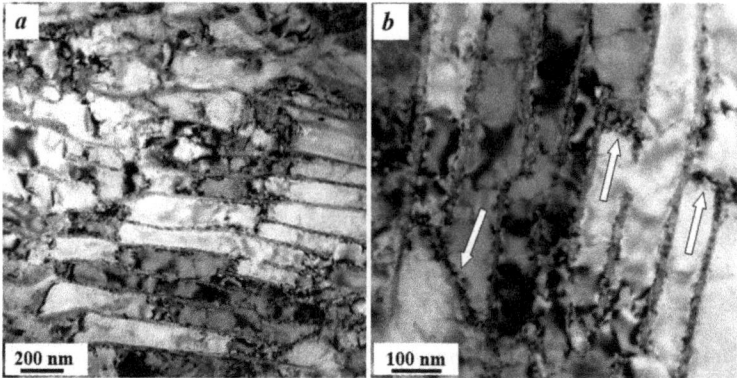

Fig. 5.14. *Electron microscopic image of a fragmented steel structure.* ε = 50%. *The arrows indicate low-angle boundaries within the ferrite plates of the pearlite colony.*

At the same time as the ferrite plates, the cementite plates are fragmented (Fig. 5.15). The size of these fragments varies from 15 to 20nm and depends weakly upon the degree of steel deformation.

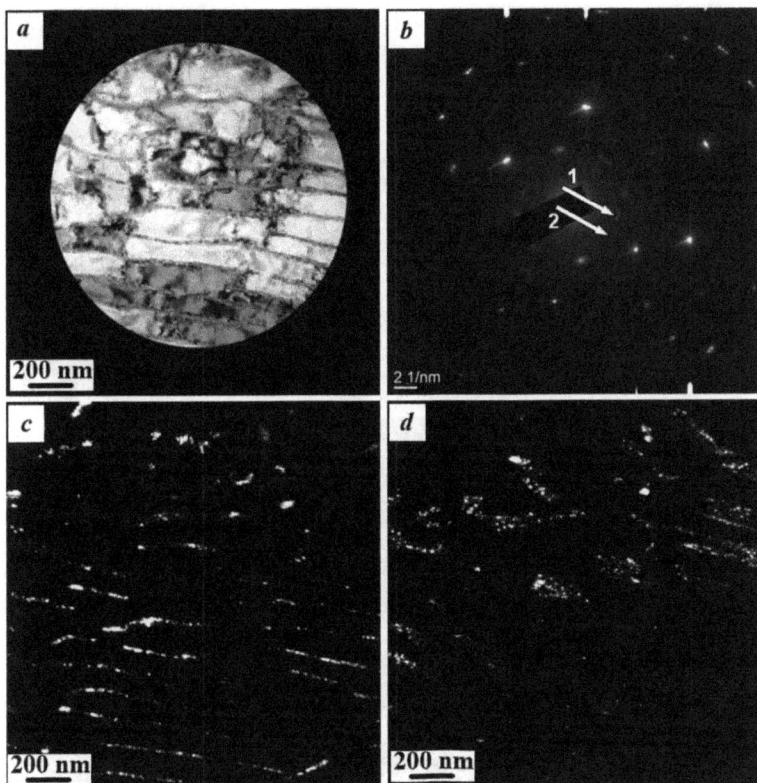

Fig. 5.15. *Electron microscopic image of a steel structure deformed at* $\varepsilon = 50\%$; *a – light field; b – electron-microscopic diffraction pattern obtained from the foil section, the image of which is shown in (a); c, d – dark field images obtained using the reflections of cementite [121] Fe₃C (c) and [211] Fe₃C (d). In (b) the arrows indicate the reflections for which dark fields are obtained, 1 – (c), 2 – (d).*

At the same time as fragmentation a dissolution and cutting of cementite plates is observed. Carbon atoms which are transferred from the cementite crystal lattice to dislocations are carried into the interplate space and form particles of tertiary cementite (Fig. 5.16*b*). The size of such particles is 2 to 4nm.

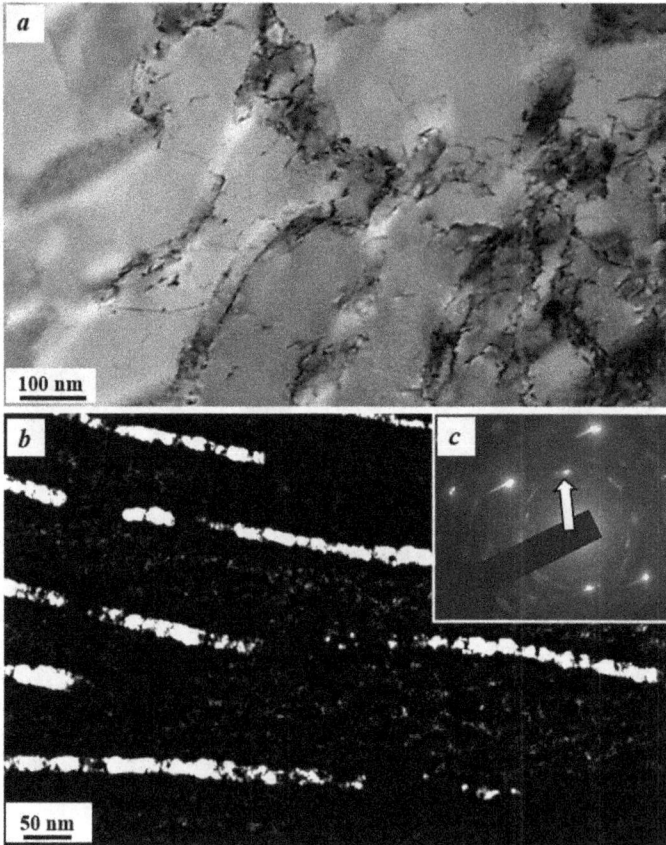

Fig. 5.16. *Electron microscopic image of a steel structure deformed at ε = 50%; a – light field; b – dark field obtained for the reflections of [012] Fe₃C + [110]α-Fe; c – electron-microscopic diffraction pattern. In (b) the arrow indicates the reflection for which the dark field is obtained.*

The deformation of pearlite grains is accompanied by a transformation of the dislocation sub-structure. In the initial structure of the steel, dislocations are distributed quasi-uniformly over the volume of ferrite plates (Fig. 5.17a). Deformation of the steel leads to the formation of dislocation-clusters around cementite particles (Fig. 5.17*b*).

Materials Research Forum LLC
https://doi.org/10.21741/9781644902776

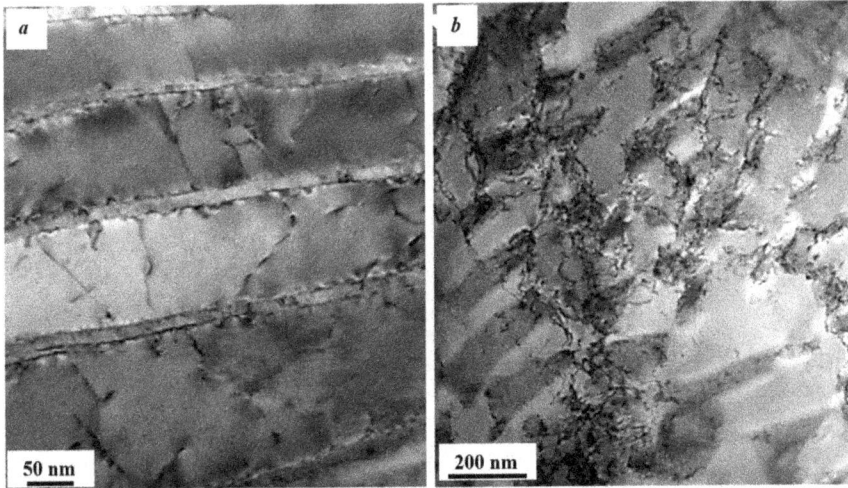

Fig. 5.17. *Electron microscopic image of the dislocation sub-structure of steel in the initial state (a) and after compression to a strain of ε = 50% (b).*

The most frequently used quantitative characteristic of the dislocation sub-structure is the scalar dislocation density <ρ>. As a result of the tests performed it was found that an increase in the degree of deformation is accompanied by a decrease in the scalar density of dislocations located within the volume of fragments (Fig. 5.18). This may be due to the incorporation of the dislocation into low-angle boundaries, as well as due to their annihilation. A similar change in the dislocation sub-structure in fragments formed during deformation was observed in [11–16].

One can estimate the density of dislocations which form the low-angle boundaries of fragments. It is shown [17, 18] that it is possible to estimate the density of dislocations in low-angle boundaries from the misorientation angle of the boundary by using the expression

$$\rho_b = 2\Theta/(bd), \tag{5.2}$$

where Θ is the misorientation angle between fragments, b is the Burgers vector of the dislocations in the low-angle boundary and d is the average size of the fragment.

The azimuthal component of the total misorientation angle can be determined from the corresponding electron-microscopic diffraction pattern [18], based upon the ratio,

$$\Theta = \Delta/R, \text{ rad}, \tag{5.3}$$

where Δ is the maximum diffraction blur (Fig. 5.19*b*) and R is the length of the radius vector of this reflection (Fig. 5.19*b*).

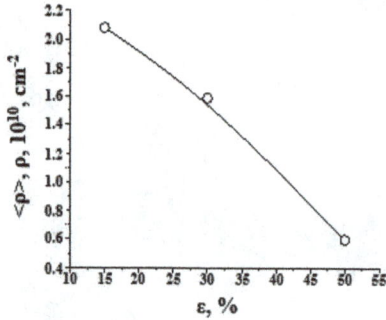

Fig. 5.18. *Dependence upon the degree of deformation of the scalar density of dislocations distributed within the bulk of fragments.*

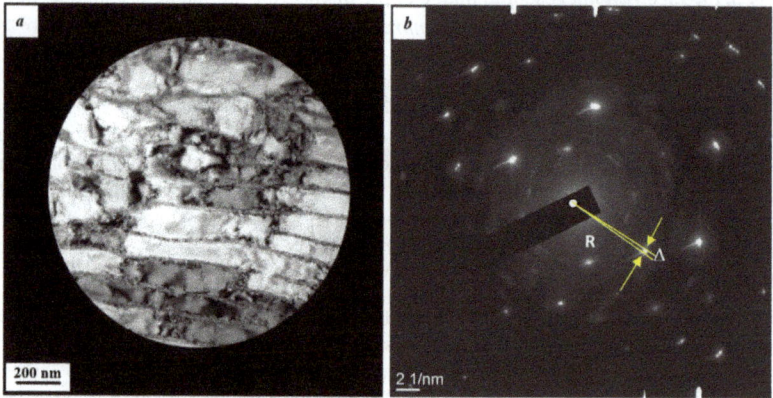

Fig. 5.19. *Electron microscopic image of a steel structure (a), b – electron-microscopic-diffraction pattern obtained from this section of the foil; the arrows indicate the reflection via which the azimuthal component of the angle of complete misorientation of the steel structure was determined.*

Taking b = 0.25nm, and the average size of the fragments to be d = 200nm, one obtains:

$\rho_b = 2\Theta/(bd) = 0.002 \text{nm}^{-2} = 0.002 \cdot \text{x } 10^{14} \text{cm}^{-2} = 2.0 \cdot \text{x } 10^{11} \text{cm}^{-2}$.

Considering that, at $\varepsilon = 50\%$, the relative content of pearlite with a fragmented structure is 0.37, one finally deduces the density of dislocations concentrated within the low-angle boundaries of the studied fragments to be,

$$\rho_b = 0.37 \times 2.0 \times 10^{11} \text{cm}^{-2} = 7.4 \times 10^{10} \text{cm}^{-2}.$$

It is possible, with a certain degree of caution, to consider this to be the density of dislocations in steel at $\varepsilon = 50\%$; but this is no longer a scalar density of dislocations.

As noted above the deformation of steel is accompanied not only by the fragmentation of pearlite but also leads to the formation of clusters of dislocations around cementite particles (Fig. 5.17b). The density of dislocations in cementite particles is estimated from the image of the structure shown in Fig. 5.20.

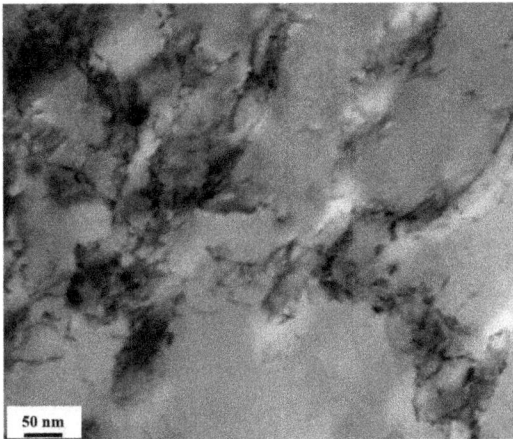

Fig. 5.20. Electron microscopic image of a dislocation sub-structure formed near to cementite particles in steel subjected to plastic deformation by uniaxial compression to $\varepsilon = 50\%$.

One can use the well-known expression [18] to determine the scalar density of dislocations,

$$< \rho > = \frac{M}{t}\left(\frac{N_1}{L_1} + \frac{N_2}{L_2}\right), \qquad (5.4)$$

where M is the magnification of the photograph of the structure, t is the thickness of the foil, N_1 and N_2 are the numbers of intersections of dislocation lines by the secant line and L_1 and L_2 are the lengths of randomly thrown secant lines along and across the selected image of the structure.

For the steel structure shown in Fig. 5.20, $M = 44$ x 10^4 and $t = 200$ x 10^{-7}cm, giving,

$$<\rho> = 9.8 \cdot 10^{10} \text{ cm}^{-2}.$$

Given that the content of pearlite having a fragmented structure is 0.37, one can assume that the remainder of the steel structure (0.63) is occupied by particles which are surrounded by dislocations, as in Fig. 5.20. In this case the scalar density of dislocations in steel at $\varepsilon = 50$ % is

$$<\rho> = 0.63 \text{ x} \cdot 9.8 \cdot \text{x } 10^{10} \text{cm}^{-2} = 6.2 \cdot \text{x } 10^{10} \text{cm}^{-2}.$$

Transmission electron microscopy of the defect sub-structure of deformed steel revealed bend extinction contours in electron-microscopic images of the steel structure (Fig. 5.21). The presence of bend extinction contours in electron-microscopic images of the structure indicates bending-torsion of the crystal lattice of this region of the material and, consequently, internal stress-fields which are bending the thin foil and accordingly hardening the material [19–21]. By analyzing the flexural extinction contours it is possible to identify the sources of internal stress-fields; i.e., to identify stress concentrators and estimate their relative magnitudes. As a result of the research carried out in this work it was found that the sources of internal stress fields are the boundaries of grains and colonies of pearlite (Fig. 5.19a), cementite plates in pearlite grains (Fig. 5.21*b*) and particles of second phase located within the bulk of ferrite plates (Fig. 5.21*b*).

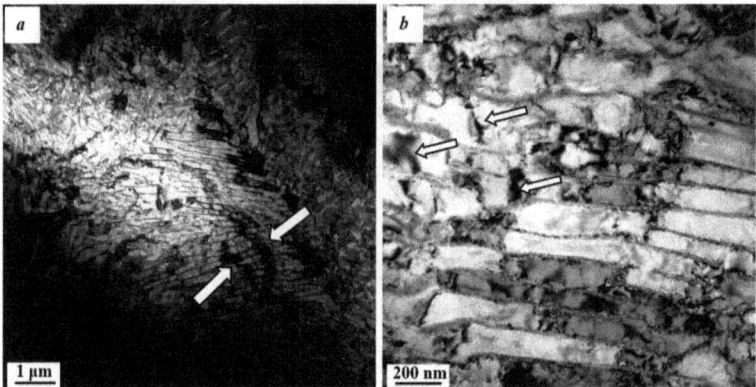

Fig. 5.21. *Structure of deformed rail steel ($\varepsilon = 50\%$). The arrows indicate extinction bend contours.*

One of the characteristics of torsion curvature of a crystal lattice is the excessive density of dislocations. Using the method of analysis of the torsion curvature of a crystal lattice, first described in [22, 23], in this paper the magnitude of the excess dislocation density was estimated. The results of the evaluation showed that the excess dislocation density decreases with increasing degree of steel deformation; similarly to the value of the scalar dislocation density (Fig. 5.18).

5.3. Mechanisms of pearlitic-steel hardening during compression

Based upon the results obtained in the previous section, on the structural-phase state and defect structure of rail-steel subjected to compressive strain, the magnitudes of the contributions of the main braking mechanisms of moving dislocations were estimated and the additive yield strength σ was determined. Such estimates were previously made for volumetrically and differentially hardened rails after various volumes of tonnage [24–26].

$$\sigma = \sigma_0 + \sigma_h + \sigma_{pearl} + \sigma_{or} + \sigma_c + \sqrt{\sigma_l^2 + \sigma_{hb}^2} \qquad (5.5)$$

This formula covers almost all of the contributions to deformation resistance. Here σ_o is the friction stress of dislocations in the crystal lattice of α-iron, σ_h is the solid-solution hardening of ferrite by alloying elements, σ_{pearl} is the hardening due to pearlite, σ_{or} is due to hardening by incoherent particles with dislocation bypassing via the Orowan mechanism, σ_{hb} is hardening by dislocation 'herringbone' patterns that cut gliding dislocations and σ_l is hardening due to internal long-range stress fields. Assessments of the contributions made by the hardening mechanisms were carried out according to the formulae given in the previous chapters.

The friction stress of dislocations in the crystal lattice of α-iron is $\sigma_o = 35\text{MPa}$ [27, 28].

Hardening of ferrite-based solid solution by the atoms of alloying elements is determined by the ratio [26]:

$$\sigma_s = \sum_{i=1}^{m} k_i C_i, \qquad (5.6)$$

where k_i is the contribution coefficient representing the increase in strength of the material at the yield point when 1% of the alloying element is dissolved in it, and C_i is the concentration of the i–th element. The i-th element can be Mn, Si, Cr, Ni, Mo, Al, P, V, Ti or Cu in the quantities available at this moment in the α-solid solution.

The hardening due to pearlite is determined by the ratio [26]:

$$\sigma_{pearl} = k_h (4.75r)^{-1/2} P_V, \qquad (5.7)$$

where P_V is the volume fraction of pearlite, r is the distance between the particles of Fe_3C and $k_h = 1.5 - 2.5 = 2\text{kGfmm}^{1/2}$.

Hardening by incoherent particles when dislocations bypass them via the Orowan mechanism is estimated from the ratio [26, 29]:

$$\sigma_{or} = B \frac{mGb}{2\pi(|r-D|)} \Phi \cdot ln\left(\left|\frac{r-D}{4b}\right|\right), \qquad (5.8)$$

where R is the average particle size, r is the distance between particle centers, Φ is a multiplying factor which depends upon the type of dislocation, B is a parameter that takes account of the uneven distribution of particles in the matrix, $\Phi = 1$, $B = 0.85$ and m is an orientation multiplier which, for bcc metals, is equal to 2.75.

Reinforcement by 'herringbone' dislocations that cut through gliding dislocations is estimated from the ratio [26, 27, 30]:

$$\sigma_{hb} = m\alpha Gb\sqrt{\rho}, \tag{5.9}$$

where m is an orientation multiplier (Schmid factor), α is a dimensionless coefficient which varies from 0.05 to 0.60 depending upon the type of dislocation pile-up (in this work, $\alpha = 0.25$), G is the shear modulus of the matrix material, b is the Burgers vector, ρ is the average value of the scalar dislocation density, $m\alpha = 1$, $G = 80000\text{MPa}$ and $b = 2.5 \cdot x \ 10^{-7}\text{mm}$.

The hardening due to internal long-range stress fields is:

$$\sigma_s = m\alpha_s Gb\sqrt{\rho_\pm} = m\alpha_s Gb\sqrt{b\chi} = \sigma_{pl} + \sigma_{el}, \tag{5.10}$$

where $\alpha_s = 0.5$ is the Strunin coefficient.

The value of the plastic component of the internal stress fields is estimated from [26, 27, 30]:

$$\sigma_{pl} = m\alpha Gb\sqrt{\rho_\pm}. \tag{5.11}$$

The value of the elastic component is estimated from [26, 27, 30]:

$$\sigma_{el} = m\alpha Gt\chi_{el}, \tag{5.12}$$

where t is the thickness of the foil, assumed to be 200nm, and χ_{el} is the elastic component of the curvature-torsion of the crystal lattice.

Analyzing the results given in Tables 5.1, 5.2 and 5.3, it can be noted that the strength of rail-steel is a multi-factor value determined by the compatible action of physical mechanisms and depends upon the degree of deformation. The main mechanisms of hardening of rail-steel during compression are long-range internal stress fields and the presence of incoherent particles of a second phase.

Table 5.1. *Quantitative parameters of steel structure in various morphological components with various degrees of plastic deformation*

Structural parameter		Pearlite			Ferrite	
		Non-fractured	Fractured	Fragmented	Non-fragmented	Fragmented
ε = 15%						
Volume fraction		70%	24%	3%	1%	2%
Transverse size of the α-phase interlayer, nm		160	120	120		
Fragment size, nm		–	–	120×400	–	400
Fe₃C	size, nm	$d = 16$	12×280	12×160		
	Volume fraction	12%	8.7%	1.5%		
Fraction of carbon		0.8%	0.6%	0.11%		
$\rho_\alpha \times 10^{-10}$, cm^{-2}		1.91	2.06	2.08	2.21	~0
$\rho_\pm \times 10^{-10}$, cm^{-2}		1.54	1.96	2.08	2.21	
$\chi = \chi_{pl} + \chi_{el}$, cm^{-1}		385	490	$650 = 520_{pl} + 30_{el}$	$1090 = 550_{pl} + 140_{el}$	$745 = 0_{pl} + 745_{el}$
ε = 30%						
Volume fraction		65%	20%	12%	0	3%
Transverse size of the α-phase interlayer, nm		160	120	120		
Fragment size, nm		–	–	120×200	–	200
Fe₃C	size, nm	$d = 18$	16×280	12×160		
	Volume fraction	12%	4.8%	0.92%		
Fraction of carbon		0.8%	0.34%	0.07%		
$\rho_\alpha \times 10^{-10}$, cm^{-2}		2.18	2.50	1.59		~0
$\rho_\pm \times 10^{-10}$, cm^{-2}		1.76	2.26	1.59		
$\chi = \chi_{pl} + \chi_{el}$, cm^{-1}		440	565	$435 = 395_{pl} + 40_h$		$745 = 0_{pl} + 745_h$

ε = 50%					
Total share		0	60%	40%	0	0
Fragment size, nm				200		
Fe₃C in the α-phase (inside fr.)	size, nm		$d = 12; r = 16$	$d = 16; r = 20$		
	Vol. fraction		1.8%	2.7%		
Share of carbon in α-phase			0.12%	0.19%		
Fe₃C in the layers of Fe₃C (on border of fr.)	size, nm		$d = 14; r = 20$	$d = 16; r = 30$		
	Vol. fraction		2.7%	1.2%		
Share of carbon			0.19%	0.09%		
$\rho_\alpha \times 10^{-10}$, cm⁻²			2.25	0		
$\rho_\pm \times 10^{-10}$, cm⁻²			2.25			
$\chi = \chi_{pl} + \chi_{el}$, cm⁻¹			$575 = 560_{pl} + 15_{el}$	$55 = 0_{pl} + 55_{el}$		

Table 5.2. *Average material parameters of the fine structure of steel following various degrees of plastic deformation*

Average structure parameter	ε = 15%	ε = 30%	ε = 50%
$\rho_\alpha \times 10^{-10}$, cm⁻²	1.92	2.11	1.35
$\rho_\pm \times 10^{-10}$, cm⁻²	1.63	1.79	1.35
$\chi = \chi_{pl} + \chi_{el}$, cm⁻¹	$425 = 410_{pl} + 15_{el}$	$470 = 445_{pl} + 25_{el}$	$365 = 335_{pl} + 30_{el}$

Table 5.3. *Contributions of various mechanisms to the hardening of the morphological components of steel, and of the overall material after various degrees of plastic deformation*

Contributions	Pearlite			Ferrite		Overall material
	Non-fractured	Fractured	Fragmented	Non-fragmented	Fragmented	
$\varepsilon = 15\%$						
Volume fraction	70%	24%	3%	1%	2%	100%
σ_{hb}, MPa	275	285	290	295	0	273
σ_{pl}, MPa	250	280	290	295	0	254
σ_{h}, MPa	0	0	40	190	1010	20
σ_{c}, MPa	–	–	550	–	350	25
σ_{0}, MPa	35	35	35	35	35	35
σ_{hard}, MPa	80	80	260	1400	1400	130
σ_{pearl}, MPa	570	250	0			460
σ_{or}, MPa			135	0	0	5
$\varepsilon = 30\%$						
Vol. fraction	65%	20%	12%	0	3%	100%
σ_{hb}, MPa	295	315	250		0	285
σ_{pl}, MPa	265	300	250		0	262
σ_{el}, MPa	0	0	55		1010	35
σ_{c}, MPa	–	–	835		750	125
σ_{0}, MPa	35	35	35		35	35
σ_{hard}, MPa	80	315	190		1400	180
σ_{pearl}, MPa	570	250	0			420
σ_{or}, MPa			135			15
$\varepsilon = 50\%$						
Vol. fraction	0	60%	40%	0	0	100%
σ_{hb}, MPa		300	0			180
σ_{pl}, MPa		300	0			180
σ_{el}, MPa		20	75			95
σ_{c}, MPa		–	750			300
σ_{0}, MPa		35	35			35
σ_{hard}, MPa		315	300			310
σ_{pearl}, MPa		250	0			150
σ_{or}, MPa		1120	645			930

The theoretical stress-strain curve obtained as a result of summing the contributions of the identified mechanisms to steel hardening agrees well with the experimental stress-strain curve (Fig. 5.22); the discrepancy being 13 to 28%. One of the reasons for the discrepancy may be the heterogeneity of the structure, as determined by the proportion of grains of lamellar pearlite and ferrite-carbide mixture. These have differing strengths and can be used to make adjustments to the deformation behavior of steel.

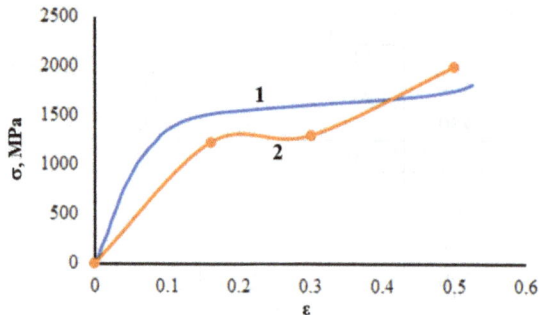

Fig. 5.22. *Experimental (1) and theoretical (2) diagrams of rail-steel deformation by uniaxial compression.*

Chapter 5 Conclusions

As a result of studies aimed at analyzing the evolution of the pearlite defect sub-structure of the lamellar morphology of P65 rail-steel samples under uniaxial compression the following was revealed:

1. The strain-hardening of rail-steel, which occurs during plastic deformation by uniaxial compression, is multi-stage in nature.

2. It is shown that the deformation is accompanied by a fragmentation of the pearlite grains, which increases with increasing degree of deformation and reaches ≈ 0.4 of the volume of the studied foil at $\varepsilon = 50\%$.

3. It was found that, with increasing degree of deformation, the average size of ferrite plate fragments decreases from 240nm ($\varepsilon = 15\%$) to 200nm ($\varepsilon = 50\%$).

4. Fragmentation of cementite plates was observed, with the size of the fragments ranging from 15 to 20nm and depending weakly upon the degree of deformation of the steel.

5. Fracture of cementite plates was detected, proceeding via their dissolution and cutting by mobile dislocations. It is shown that carbon atoms which are transferred from the cementite lattice to dislocations are carried into the inter-plate space and form particles of tertiary cementite with sizes of 2 to 4nm.

6. The formation of an inhomogeneous dislocation sub-structure was revealed, caused by the impeding of dislocations by cementite particles during steel deformation.

7. It was found that an increase in the degree of deformation is accompanied by a decrease in the scalar and excess dislocation density, due possibly to the drifting of dislocations into low-angle boundaries, or to their annihilation.

The main hardening factor of the studied rail-steel in the initial stage of deformation by uniaxial compression ($\varepsilon \approx 15\%$) is thus the presence of grains of lamellar pearlite. With increasing degree of deformation, the role played by this factor decreases due to the fracture of cementite plates. The role of the contributions to steel hardening played by the formation of a solid solution (due to cementite dissolution), fragmentation (due to a decrease in the size of fragments and an increase in the relative content of the fragmented structure) and incoherent particles of the carbide-phase increases with increasing degree of steel deformation. In spite of the heterogeneity of the structure, the agreement between experimentally measured and calculated values of steel strength is quite good.

Chapter 5 References

[1] Fractography and Atlas of Fractograms (Translated from English), Ed. J.Fellows, Metallurgy, 1982, 489pp.

[2] Gromov, V.E., Ivanov, Y.F., Yuryev, A.A., Morozov, K.V., Konovalov, S.V., Differentially hardened rails: evolution of structure and properties during long-term operation, SibSIU Publishing Center, Novokuznetsk, 2017, 197pp.

[3] Yuryev, A.A., Kuznetsov, R.V., Gromov, V.E., Ivanov, Y.F., Shlyarova, Yu.A., Long Rails: Structure and Properties after Long-Term Operation, Polygraphist, Novokuznetsk, 2022, 311pp.

[4] Yuriev, A.A., Gromov, V.E., Ivanov, Y.F., Rubannikova, Y.A., Starostenkov, M.D., Tabakov, P.Y., Structure and Properties of Long Rails after Extreme Long-Term Operation, Materials Research Forum, LLC, 2021, 185pp. https://doi.org/10.21741/9781644901472

[5] Tushinsky, L.I., Bataev, A.A., Tikhomirova, L.B., Pearlite Structure and Structural Strength of Steel, VO Nauka, Novosibirsk, 1993, 280pp.

[6] Schastlivtsev, V.M., Mirzaev, A.A., Yakovleva, I.L., Okishev, K.Y., Tabatchikova, T.I., Khlebnikova, Y.V., Pearlite in Carbon Steels, Ural Branch of the Russian Academy of Sciences, Ekaterinburg, 2006, 312pp.

[7] Ivanov, Y.F., Gromov, V.E., Popova, N.A., Konovalov, S.V., Koneva, N.A., Structural-Phase State and Mechanisms of Hardening of Deformed Steel, Polygraphist, Novokuznetsk, 2016, 510pp.

[8] Trefilov, V.I., Moiseev, V.F., Pechkovsky, E.P., Deformation Hardening and Destruction of Polycrystalline Metals, Naukova Dumka, Kiev, 1989, 256pp

[9] Koneva, N.A., The nature of stages of plastic deformation, Soros Educational Journal, 1998, 10, 99-105.

[10] Podrezov, Y.N., Firstov, S.A., Two approaches to the analysis of hardening stress-strain curves, Physics and Technology of High Pressures, 2006, 16[4] 37-48.

[11] Gromov, V.E., Kozlov, E.V., Bazaykin, V.I. et al., Physics and Mechanics of Drawing and Volumetric Stamping, Nedra, 1997, 293pp.

[12] Kozlov, E.V., Popova, N.A., Ignatenko, L.N., Teplyakova, L.A., Klopotov, A.A., Koneva, N.A., Influence of substructure type on carbon redistribution in martensitic steel during plastic deformation, Izv. Universities - Physics, 2002, 45[3] 72-86. https://doi.org/10.1023/A:1020396717448

[13] Kozlov, E.V., Popova, N.A., Koneva, N.A., Fragmented substructure formed in bcc steels during deformation, Izvestiya RAS, 2004, 68[10] 1419-1428.

[14] Kozlov, E.V., Popova, N.A., Koneva, N.A., Dimensional effect in dislocation substructures of metallic materials, Fundamental Problems of Modern Metal Materials, 2009, 6[2] 14-24.

[15] Koneva, N.A., Kozlov, E.V., Popova, N.A., Influence of grain size and fragments on dislocation density in metallic materials, Fundamental Problems of Modern Materials Science, 2010, 7[1] 64-70.

[16] Kozlov, E.V., Popova, N.A., Koneva, N.A., Scalar density of dislocations in fragments with different types of substructures, Letters About Materials, 2011, 1 15-18. https://doi.org/10.22226/2410-3535-2011-1-15-18

[17] Norstrom, I.A., On the yield strength of quenched low-alloy lath martensite, Scandinavian J.Met., 1976, 5[4] 159-165.

[18] Utevsky, L.M., Diffraction Electron Microscopy in Metal Science, Metallurgy, 1973, 584pp.

[19] Gromov, V.E., Yuriev, A.B., Morozov, K.V., Ivanov, Y.F., Microstructure of Quenched Rails, ISP, Cambridge, 2016, 153pp.

[20] Ivanov, Yu.F., Kornet, E.V., Kozlov, E.V., Gromov, V.E., Hardened Structural Steel: Structure and Mechanisms of Hardening, SibSIU Publishing House, Novokuznetsk, 2010, 174pp.

[21] Ivanov, Yu.F., Gromov, V.E., Popova, N.A., Konovalov, S.V., Koneva, N.A., Structural-Phase States and Mechanisms of Hardening of Deformed Steel, Polygraphist, Novokuznetsk, 2016, 519pp.

[22] Koneva, N.A., Lychagin, D.V., Teplyakova, L.A., Kozlov, E.V., Dislocation-disclination substructures and hardening, Theoretical and Experimental Study of Disclinations, FTI, 1984, 116-126.

[23] Koneva, N.A., Kozlov, E.V., Trishkina, L.I., Lychagin, D.V., Long-range stress

fields, curvature-torsion of the crystal lattice and stages of plastic deformation. Measurement methods and results, New Methods in Physics and Mechanics of Deformable Solids, Proceedings of the International Conference, TSU, Tomsk, 1990, 83-93.

[24] Yuriev, A.A., Gromov, V.E., Ivanov, Yu.F., Rubannikova, Yu.A., Structure and Properties of Long-Length Differentially Hardened Rails after Extremely Long Operation, Polygraphist, Novokuznetsk, 2020, 253pp.

[25] Ivanov, Yu.F., Kormyshev, V.E., Gromov, V.E., Yuryev, A.A., Glezer, A.M., Rubannikova, Yu.A., Mechanisms of hardening of metal rails during long-term operation, Questions of Materials Science, 2020, 3, 17-28. https://doi.org/10.22349/1994-6716-2020-103-3-17-28

[26] Yuriev, A.A., Gromov, V.E., Ivanov, Yu.F., Rubannikova, Yu.A., Starostenkov, M.D., Tabakov, P.Y. Structure and Properties of Long Rails after Extreme Long-Term Operation, Materials Research Forum, LLC, 2021, 193pp. https://doi.org/10.21741/9781644901472

[27] Ivanov, Y.F., Kornet, E.V., Kozlov, E.V., Gromov, V.E., Hardened Structural Steel: Structure and Mechanisms of Hardening, SibSIU Publishing House, Novokuznetsk, 2010, 174pp.

[28] Ivanov, Y.F., Gromov, V.E., Nikitina, E.N., Bainitic Constructional Steel, Structure and Hardening Mechanisms, Cambridge, 2017, 121pp.

[29] Yao M.J., Welsch, E., Ponge, D., Haghighat, S.M.H., Sandlöbes, S., Choi, P., Herbig, M., Bleskov, I., Hickel, T., Lipinska-Chwalek, M., Shantraj, P., Scheu, C., Zaefferer, S., Gault, B., Raabe, D., Strengthening and strain hardening mechanisms in a precipitation-hardened high-Mn lightweight steel, Acta Materialia, 2017, 140, 258-273. https://doi.org/10.1016/j.actamat.2017.08.049

[30] Koneva, N.A., Kiseleva, S.F., Popova, N.A., Evolution of Structure and Internal Stress Fields, Austenitic Steel, Laplambert Academic Publishing, Saarbrucken, 2017, 145pp.

Overall Conclusion

Strain-hardening results from the formation of internal stress fields caused by plastic flow, which is generated by the evolution of an ensemble of interacting dislocations. Despite the fact that dislocation theory has been used for more than 50 years to explain the processes of strain-hardening, it remains one of the most practically important and most intensively developing areas in the physics of steels. The physics of strain-hardening during the plastic deformation of steels having various structures is based upon the dynamics of plastic deformation and involves such physical parameters as the velocity of dislocations, their density and their multiplication. Strain-hardening significantly affects the resistance of steels to large plastic deformations and fracture. The physics of strain-hardening developed historically in two interrelated directions, taking account both of the motion of dislocations and of their multiplication. These physical characteristics and the characteristics closely related to them (dislocation lifetime, yield strength, etc.) are considered to be significant parameters in the plastic deformation of steels. The most important stage in the creation of the microscopic theory of strain-hardening is the analysis of various types of interaction between the ensemble of dislocations and the defect sub-structure of steels of various classes.

A number of fundamental circumstances indicate that the physics of strain-hardening is still far from being complete. One of the main factors that will undoubtedly shape this science is the increased interest in materials having submicro- and nanocrystalline structures and in high-entropy alloys.

www.ingramcontent.com/pod-product-compliance
Lightning Source LLC
Chambersburg PA
CBHW071702210326
41597CB00017B/2297